Using Economic Incentives to Regulate Toxic Substances

W0112892

Molly K. Macauley, Michael D. Bowes, and Karen L. Palmer

RFF PRESS
RESOURCES FOR THE FUTURE

First published in 1992
by Resources for the Future, Inc.

This edition first published in 2016 by Routledge
2 Park Square, Milton Park, Abingdon, Oxon, OX14 4RN
and by Routledge
711 Third Avenue, New York, NY 10017

Routledge is an imprint of the Taylor & Francis Group, an informa business

Publisher's Note
The publisher has gone to great lengths to ensure the quality of this reprint but
points out that some imperfections in the original copies may be apparent.

Disclaimer
The publisher has made every effort to trace copyright holders and welcomes
correspondence from those they have been unable to contact.

A Library of Congress record exists under LC control number: 92039781

ISBN 13: 978-1-138-95656-8 (hbk)
ISBN 13: 978-1-315-66568-9 (ebk)
ISBN 13: 978-1-138-95658-2 (pbk)

Routledge Revivals

Using Economic Incentives to Regulate Toxic Substances

Using case studies, the authors evaluate the potential attractiveness of incentive-based policies for the regulation of four specific toxic substances: chlorinated solvents, formaldehyde, cadmium, and brominated flame retardants. Originally published in 1992, the authors provide a compelling demonstration of the role of case studies in determining the appropriate regulatory approach for the specific toxic substances. This is a valuable title for students concerned with environmental issues and policy making.

USING ECONOMIC INCENTIVES TO REGULATE TOXIC SUBSTANCES

Molly K. Macauley
Michael D. Bowes
Karen L. Palmer

Resources for the Future
Washington, D.C.

0-915707-65-9/92

Printed in the United States of America

Published by Resources for the Future
1616 P Street, NW, Washington, DC 20036

Library of Congress Cataloging-in-Publication Data

Macauley, Molly K.
   Using economic incentives to regulate toxic substances / Molly K. Macauley, Michael D. Bowes, Karen L. Palmer.
      p.   cm.
   Includes bibliographical references and index.
   ISBN 0-915707-65-9
   1. Hazardous substances—Government policy—United States.
   2. Hazardous substances—Law and legislation—United States.
   3. Hazardous wastes—Government policy—United States.   4. Hazardous wastes—Law and legislation—United States.   5. Incentives in industry—United States.   6. Tax incentives—United States.
   I. Bowes, Michael D. II. Palmer, Karen L. III. Title.
   HC110.P55M32   1992
   363.72'87'0973—dc20   92-39781
                                                                             CIP

This book is the product of RFF's Energy and Natural Resources Division, Douglas R. Bohi, director, and of the Quality of the Environment Division, Raymond J. Kopp, director. It was edited and indexed by Julie Phillips; the cover was designed by Kelly Design.

∞The paper in this book meets the guidelines for permanence and durability of the Committee on Production Guidelines for Book Longevity of the Council on Library Resources.

Contents

List of Figures and Tables

FIGURES

Foreword

Since the mid- to late 1960s, economists at Resources for the Future and elsewhere have sounded a common theme when discussing environmental regulation. Specifically, they have recommended that, wherever possible, so-called command-and-control regulation (for instance, requirements that manufacturers install specific types of pollution control equipment) be replaced by the use of economic incentives such as the imposition of taxes on pollutant emissions or the introduction of a system of marketable permits limiting the amount of pollution that can be discharged during some specified period of time. In demonstrating the considerable advantages of incentive-based approaches—most importantly, the cost savings they make possible—environmental economists have almost always used as examples air and water pollutants that are discharged from easily identifiable smokestacks or outfall pipes at which continuous monitoring of emissions is at least conceivable if not already currently practiced.

There is nothing wrong with these examples at all. In fact, because much work had been done on possible incentive-based approaches to air pollution control, it was possible in the 1990 amendments to the Clean Air Act to include a dramatically innovative system of marketable "pollution allowances" to implement the required reductions in sulfur dioxide emissions from electric utility generating stations. Nevertheless, not all environmental regulatory problems are as conceptually straightforward as the canonical air or water pollution control examples in textbooks. In addition to traditional air and water pollutants, the U.S. Environmental Protection Agency (EPA) must regulate the manufacture and use of chemicals and other toxic substances that can generate risks to human health and the environment not as a result of emissions from discrete smokestacks or outfall pipes but rather from a wide variety of sometimes very diverse applications in a whole host of different locations. This raises an obvious question: can incentive-based approaches be used in the generally very different conditions under which toxic substances are produced and used, and—if so—will they have the same resource-conserving and other desirable features? This is the question Molly Macauley, Michael Bowes, and Karen Palmer set out to answer in this book.

In addressing this question, they depart from much of the literature on incentive-based approaches in another (I think refreshing) respect. That is, they couch their entire analysis in the context of four case studies, each of which involves a particular type of substance that the EPA must actually regulate under the Toxic Substances Control Act of 1976 (TSCA). No abstract theorizing for them. Rather, they immerse themselves in the real-world conditions under which chlorinated solvents, formaldehyde, cadmium, and brominated flame retardants are produced and/or used in the United States; the respects in which this production and use generates risks to human health and/or the environment; and the nature of the incentive-based approaches that might be used to help eliminate or at least reduce these risks.

In the course of their analysis, Macauley, Bowes, and Palmer consider a rich menu of incentives. These include the pollution taxes and marketable permits that have been the mainstays of the economists' armamentarium regarding conventional air and water pollution problems. But they range beyond these to consider the provision of information to workers or consumers through labeling and other means, the adoption of deposit-refund schemes, the targeting of enforcement efforts, and other approaches as well. In each case, they try to match the characteristics of the incentive-based approach to the special real-world characteristics of the chemical substance and use in question. In this regard, then, they do a service not only for economists interested in the possible applications of their discipline to environmental policy, but also for businessmen and women, environmentalists, government regulators, and others interested in the more efficient and effective regulation of an industry that is of great importance to the United States.

I would like to believe that earlier RFF books about air and water pollution control have helped pave the way for a revolutionary and more enlightened approach to environmental regulation now finding its way into law. I believe *Using Economic Incentives to Regulate Toxic Substances* has the potential to revolutionize an important and relatively new area of environmental regulation—the control of chemicals under TSCA—although I trust I may be excused for hoping *this* revolution takes less than 30 or so years!

PAUL R. PORTNEY
VICE PRESIDENT

OCTOBER 1992   RESOURCES FOR THE FUTURE

Acknowledgments

The research on which this book is based was sponsored by the Office of Toxic Substances, U.S. Environmental Protection Agency (EPA), and Resources for the Future (RFF). We would like to thank Christine Augustyniak, the project officer at EPA during early stages of the report, and Paul R. Portney, RFF's vice president, for financial support from their respective organizations and for their intellectual support. Barbara S. Glenn served as our consultant on understanding the chemistry of toxic substances.

Nishkam S. Agarwal, the project officer at EPA during final stages of the report, provided extensive direction, and many of his comments are included verbatim in this study.

Numerous additional individuals have given us helpful advice and comments, including Robert Lee II and Jean E. Parker, EPA; Richard P. Theroux, Office of Management and Budget; and Frank S. Arnold and Frances G. Sussman, ICF Consulting Associates, Inc. A variety of experts in industry and at other organizations have patiently shared observations about the chemical industry in general or contributed hours of knowledge about the specific toxic substances we consider in this research. These include Karen L. Florini, Environmental Defense Fund; John F. Murray, Formaldehyde Institute; William H. McCredie, National Particleboard Association; Stephen P. Risotto, Halogenated Solvents Industry Alliance; Cheryl O. Morton, Synthetic Organic Chemical Manufacturers Association, Inc.; Katy A. Wolf, Institute for Research and Technical Assistance; Brenda D. Pulley, National Association of Chemical Recyclers; and Douglas G. Bannerman and Frederick E. Nicholson, National Electrical Manufacturers Association.

We also thank several anonymous reviewers recruited between the first and final drafts of the study. We incorporated many of their suggestions, to the substantial benefit of the study.

None of the individuals mentioned above necessarily endorses our conclusions; indeed, responsibility for errors, omissions, and conclusions rests wholly with the authors.

We also are deeply grateful to Diane DeWitt, John Powers, and Gayle Killam for excellent research assistance. Kay Murphy very adeptly managed the production of the manuscript in its final pre-publication form. And, finally, we are especially grateful for the care, patience, and efficiency with which Dorothy Sawicki managed the editing and publication of this book. Her professionalism made a significant contribution to the product.

1

Introduction

More than 60,000 chemicals (excluding pharmaceuticals or pesticides) enter into the many products and services that shape today's life-styles. Taken together, these chemicals comprise a huge industry—in the United States alone, sales during 1990 were over $200 billion. The sheer variety, ubiquity, and economic importance of chemicals means that effective regulation to safeguard against undesirable health or environmental side effects is quite challenging.

Traditionally, regulation to bring about these safeguards has taken the form of "command and control"—that is, banning or restricting the production or use of a chemical, requiring the substance to be reformulated, mandating recycling of the substance when exposure occurs during disposal, or otherwise restricting its use and distribution.[1] For example, the Toxic Substances Control Act of 1976 (TSCA) prohibits the manufacture, processing, distribution in commerce, and use of polychlorinated biphenyls (PCBs) because of evidence that they pose a significant risk to public health and the environment.[2] Similarly under TSCA, most uses of chlorofluorocarbons (CFCs) as aerosol propellants have been banned because CFCs may deplete the stratospheric ozone layer, which could lead to an increase in skin cancer, cataracts, and other adverse health and ecological effects.

TSCA is almost unique among environmental regulations in requiring that economic considerations be explicitly included in any actions taken under the act. Among the other six major statutes under

---

[1]Typically in the literature on environmental regulation, "command and control" refers to the regulator's specification of control technologies or performance standards that producers must adopt. In the case of toxic substances, command and control has most frequently taken the form of output or other mandated restrictions such as those described above. See Schultze (1977) for one of the seminal discussions of command and control in the context of environmental regulation.

[2]By the early 1990s, the riskiness of the toxicity of PCBs was being reevaluated.

1

which the Environmental Protection Agency (EPA) regulates, only the Federal Insecticide, Fungicide, and Rodenticide Act of 1972 features a major role for benefit–cost balancing in setting standards. Specifically, section 6 of TSCA mandates that once an "unreasonable risk of injury to health or the environment" has been established,[3] then the "least burdensome" form of regulation must be selected:

(a) Scope of Regulation.—If the Administrator finds that there is a reasonable basis to conclude that the manufacture, processing, distribution in commerce, use, or disposal of a chemical substance or mixture, or that any combination of such activities, presents or will present an unreasonable risk of injury to health or the environment, the Administrator shall by rule apply one or more of the following requirements to such substance or mixture to the extent necessary to protect adequately against such risk using the *least burdensome* requirements [emphasis added].

However, TSCA has been infrequently applied for a variety of reasons, including the difficulty in defining and measuring unreasonable risk and the extensive legislation that overlaps TSCA (for example, regulation of occupational health, waste management, consumer product safety—see figure 1-1). The most recent headline-making legal challenge concerning the "least burdensome" provision occurred in October 1991. At that time the U.S. Fifth Circuit Court of Appeals in *Corrosion Proof Fittings* v. *Environmental Protection Agency* found that EPA had failed to satisfy the provision in banning asbestos from various products. The court noted:

While the EPA may have shown that a world with a complete ban of asbestos might be preferable to one in which

---

[3]TSCA does not explicitly define unreasonable risk, but as Shapiro (1990, p. 211) notes, the U.S. House of Representatives' report accompanying the legislation gives some guidance: "In general, a determination that a risk associated with a chemical substance or mixture is unreasonable involves balancing the probability that harm will occur and the magnitude and severity of that harm against the effect of the proposed regulatory action on the availability to society of the benefits of the substance or mixture, taking into account the availability of substitutes for the substance or mixture which do not require regulation, and other adverse effects which such proposed action may have on society. . . . The committee [House Committee on Interstate and Foreign Commerce] has limited the [EPA] Administrator to taking action only against unreasonable risks because to do otherwise assumes that a risk free society is attainable, an assumption that the committee does not make." (Toxic Substances Control Act, H. Rept. 1341 to accompany H.R. 1403, 94th Cong., 2d sess. 14–15 (1976).) See discussion in U.S. General Accounting Office (1984a, 1984b).

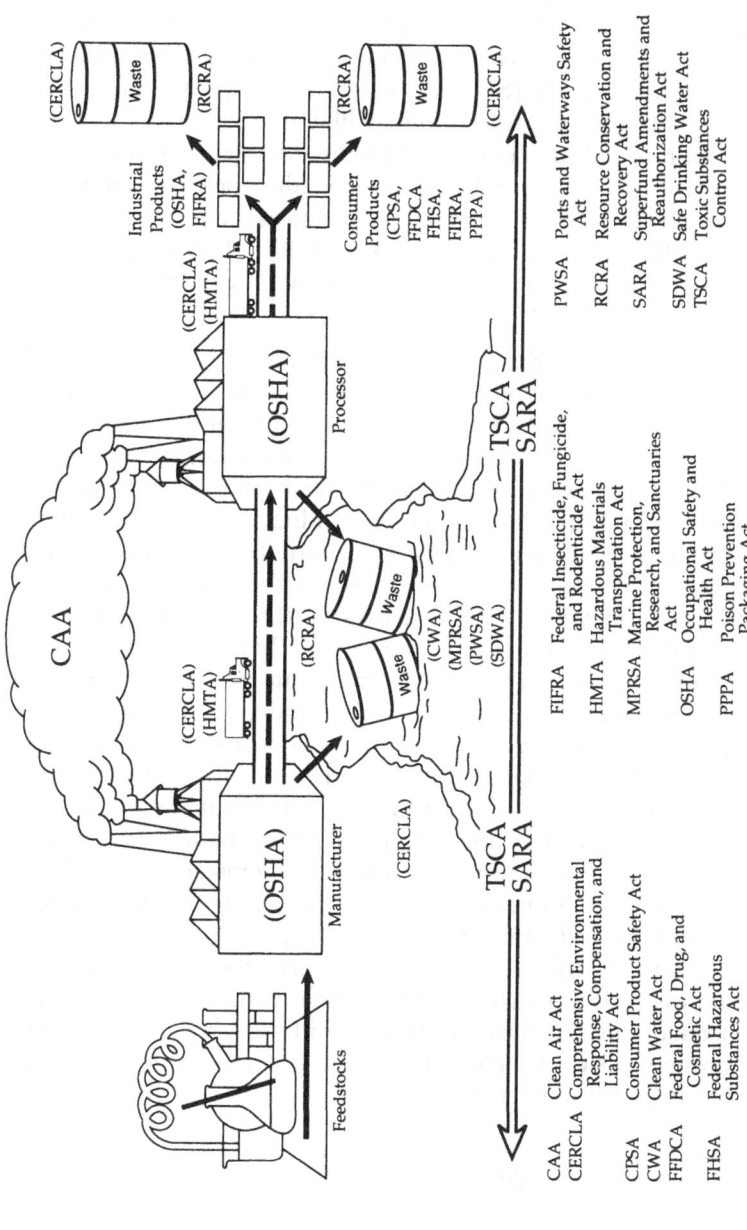

Figure 1-1. Legislative authorities affecting the life cycle of a chemical. (SARA added to figure by Macauley, Bowes, and Palmer.)

Source: Adapted from U.S. General Accounting Office, *EPA's Efforts to Identify and Control Harmful Chemicals in Use* (Washington, DC, June 13, 1984).

CAA   Clean Air Act
CERCLA Comprehensive Environmental
        Response, Compensation, and
        Liability Act
CPSA   Consumer Product Safety Act
CWA   Clean Water Act
FFDCA   Federal Food, Drug, and
        Cosmetic Act
FHSA   Federal Hazardous
        Substances Act

FIFRA   Federal Insecticide, Fungicide,
        and Rodenticide Act
HMTA   Hazardous Materials
        Transportation Act
MPRSA   Marine Protection,
        Research, and Sanctuaries
        Act
OSHA   Occupational Safety and
        Health Act
PPPA   Poison Prevention
        Packaging Act

PWSA   Ports and Waterways Safety
        Act
RCRA   Resource Conservation and
        Recovery Act
SARA   Superfund Amendments and
        Reauthorization Act
SDWA   Safe Drinking Water Act
TSCA   Toxic Substances
        Control Act

there is only the current amount of regulation, the EPA has failed to show that there is not some intermediate state of regulation that would be superior to both the currently-regulated and the completely-banned world. Without showing that asbestos regulation would be ineffective, the EPA cannot discharge its TSCA burden of showing that its regulation is the least burdensome available to it.

Upon an initial showing of product danger, the proper course for the EPA to follow is to consider each regulatory option, in the order mandated by Congress, and the costs and benefits of regulation under each option. [*Corrosion Proof Fittings* v. *Environmental Protection Agency*, No. 89-4596, slip op. at 25 (5th Cir. 1991).]

The court finding is perhaps but the most recent and direct result of a growing concern about the cost of environmental regulation. Executive Order No. 12291, issued during the Reagan administration, requires benefit–cost analysis for all proposed major governmental regulatory actions, and in 1992, President Bush called for a temporary moratorium on new federal regulations, citing their cost burden.

The issue we consider in this book is whether the menu of cost-effective options in regulating toxic chemicals should be expanded. Specifically, we ask whether there are forms of regulation—we focus on marketlike, incentive-based approaches—that might serve as desirable, cheaper alternatives to more traditional command and control. Incentive-based approaches might include, for example, taxes or tradable permits to govern the production or use of a toxic chemical, or a system of deposits and refunds to control disposal of a chemical when toxicity is a concern at the disposal stage. Another possibility is product labeling to provide information to users to encourage safer use of, or substitution for, chemicals to which exposure is a concern primarily for the individual user. We consider these approaches, and some additional strategies, in the case studies presented in this book. In each case, we assume that the goal of intervention is to internalize or capture cost-effectively the externalities associated with chemical production and use—that is, the health or environmental side effects that are not generally reflected in the market prices of these chemicals.

OUR APPROACH

We draw on a large and growing economic literature (see the section on Relevant Literature, below) to suggest how incentive-based approaches to regulation can be designed for four substances: chlorinated solvents, formaldehyde, cadmium, and brominated flame retardants (BFRs). We chose these substances for review because they seemed to call for different approaches and would thus illustrate a variety of applications of incentive mechanisms, and also because they are currently of concern to regulators.[4] Although we do not estimate the magnitude of the potential cost savings to society of incentive-based approaches over command and control (an important topic for future research), we do suggest some of the key factors that will affect the size of the savings.

This study takes as a given that the substances we consider are to be regulated. In all four case studies, we note but do not review the ongoing research bearing on whether they present unreasonable risk to health or the environment. In other words, although we discuss some of the risks associated with each substance and hence allude to some of the benefits of controlling them, we do *not* make benefit–cost comparisons here. Not only are such comparisons useful to draw in general; as indicated above, they are required under TSCA.

Here, however, we assume that the policymaker has decided to regulate the substances, and take it as our task to explore alternatives to command and control. For each of the four examples, we first study the nature of exposure—whether it is to humans or the environment; whether it is to the user of the substance/product alone, or to society more generally; and whether exposure occurs through contact with air, water, or soil. Second, we examine the stages of the product life cycle (that is, from feedstock to manufacture, distribution, final use, and disposal) during which exposure may occur. We then discuss the heterogeneity of uses and other market factors and the incentive-based approaches that seem appropriate given these characteristics. In chapters 2 through 5, we proceed from the case study methodology that presents our review of the characteristics of each substance to a final section on the regulatory approaches that we recommend.

Although our research leads us to believe that toxic substance regulation will inevitably require the tailoring of approaches to accommodate specifics of chemicals and their industries, in the concluding chapter we present some overarching generalizations about incentive-based approaches. Accordingly, in that chapter, we draw inferences from the other direction—that is, identifying which char-

---

[4]The substances are under study by regulators in the United States, in committees of the Organisation for Economic Co-operation and Development, or both.

acteristics of various incentive-based approaches match up well against the characteristics of chemical substances and why.

GENERAL OBSERVATIONS

Several of the characteristics of toxic substances that come up throughout this book suggest that regulating these substances may well require approaches different from those suitable for the typical pollution problems addressed in the existing literature on incentive-based strategies. The textbook air and water pollution problems generally involve homogeneous pollutants associated with one stage of production at a readily identifiable source.[5] The canonical example might be sulfur dioxide emissions from smokestacks of coal-fired or oil-fired electric power plants. Incentive-based strategies, such as tradable permits, have been adopted in this case to allow utilities with different costs of emissions control to minimize these costs of meeting a national sulfur dioxide emissions limit. In contrast to this example, three characteristics of toxic substances figure prominently in our analysis and complicate a straightforward application of the general results in the literature on incentive strategies for environmental regulation.[6] The characteristics are these:

1. Risks occur at varying stages of the life cycle. The potential for risks to health and the environment may occur at the mine mouth or during production of the feedstock, during production of intermediate products that use the chemical as an input, during use by industry or households, and upon disposal. Thus, regulatory intervention to safeguard against risk may be necessary at these multiple stages and may need to take different forms (for example, a tax on intermediate production to mitigate air or water pollution as well as a label on the final product to safeguard end users).

---

[5]Extensions of the literature demonstrate that even the typical cases are not straightforward. Extensions include nonpoint pollution sources (for example, pesticide runoff into waterways), spatial variation in damages, multimedia damages, and the challenge of monitoring and enforcing compliance.

[6]Even in the utilities example, however, analysts applying the more general literature have had to address specific characteristics of the industry, such as spatial variation in damages associated with emissions (e.g., Tietenberg [1978]; Seskin, Anderson, and Reid [1983]; Baumol and Oates [1988]) and the effects of rate-based regulation on the treatment of permit costs in the utilities' decisions (Bohi and Burtraw, 1991). These examples further demonstrate the usefulness of detailed case studies in designing incentive-based strategies.

2. Distribution of exposure risk varies markedly across heterogeneous products and uses. It is not always the case that *all* products or uses of a chemical pose potential risks; nor is the nature of the risk always the same (that is, the risk may be to people or the environment, it may apply to society as a whole or merely to the end user, it may harm one or several environmental media—for example, air, water, soil). Thus, regulatory intervention may need to be targeted to safeguard against risk without unduly restricting less harmful applications—that is, it may need to be highly product- or use-specific.

3. Generally, a wide scope of substitute products or production processes exists. Regulating a substance (or one of the products in which it appears or uses to which it is put) is likely to induce substitution of another product or process. (Generally speaking, it appears to reflect the very nature of chemicals that small modifications in their makeup can provide a substitute product, although our case studies do suggest a few cases where demand appears quite inelastic, implying few substitutes.) Of course, such substitution possibilities are an intended consequence of intervention. However, the existence of potentially more harmful substitutes for a particular toxic substance or class of substances implies that toxic substance regulation, if too narrow in scope, could lead to increased levels of environmental and health risk. Consequently, toxic substance regulations must be defined broadly enough to incorporate any potentially harmful substitutes.

Taken together, these characteristics of the chemical industry lead us to the following observations with regard to regulating toxic substances:

Targeting intervention to focus on specific life-cycle stages or end uses is likely to be desirable. However, such targeting may entail significant administrative and enforcement costs. A blunt instrument, such as a tax on all production of a substance, may be administratively easy. (Typically, there are far fewer producers of chemicals upstream in the production process—at the mine mouth or feedstock level—than there are in intermediate production of products containing the chemicals, and there are far fewer producers than the thousands of diverse end users.) But such a tax will generally be passed forward, reducing use, in applications for which there are few substitutes rather than in applications for which risks of human or environmental exposure are greatest. When exposure risk is concentrated at these life-cycle stages, a tax targeted at intermediate stages of production or at certain end uses will better ensure that risks are mitigated, but may be quite costly to administer and enforce.[7]

If large enough, these costs could outweigh the benefits from an incentive-based approach targeted at the end use compared with, say, less finely differentiated incentive-based approaches taken farther upstream or even an outright ban on production of a substance.[8] This balancing of the benefits of a targeted approach against its administrative costs appears to apply in all case studies examined here. The number and diversity of producers to regulate are far fewer the farther upstream (that is, at the initial production stages—at the mine mouth or feedstock level) one goes. Moving downstream leads to a progressively larger and more diverse set of intermediate producers and, ultimately, thousands of final product users.

The most desirable intervention strategies are self-enforcing. The property of self-enforcement is clearly desirable for all types of regulation (whether command and control or incentive-based) and in all circumstances (whether a single-source, homogeneous pollutant or a multiple-source, multimedia pollutant), but this property is particularly important in the case of toxic substances, given their ubiquity and heterogeneity. Opportunities may well arise—and be easy to exploit—for evading a toxic substance tax or operating in such a manner as to elude the purchase of a permit. The large numbers of intermediate producers may make it easy, for example, to resell untaxed or nonpermitted quantities ostensibly intended for benign uses to a producer who uses the substance in a riskier process or final product. Or, in deposit-refund schemes, some chemicals may be readily and relatively undetectably diluted so as to increase refunds.

For these reasons, intervention strategies that might be self-enforcing, such as deposit-refund schemes modified to reduce counterfeiting, or taxes or permits that allow cost-effective monitoring (perhaps making use of the delivery manifests that are already required to track the distribution of many chemicals) are a focus of this approach. Strategies to penalize violators by increasing the prob-

---

[7]This issue of where to intervene in the life cycle also underlies environmental regulation of hazardous wastes. Discussion frequently focuses either on end-of-pipe control or reduction of source generation when, in fact, a comparative evaluation of the costs and benefits of intervention at *both* life-cycle stages could better ensure cost-effective regulation. Hahn (1988) takes this approach and develops a menu of regulatory options for hazardous waste using a framework similar to the one in this study.

[8]An outright ban on feedstock production of a chemical will also preclude even those uses that might be less risky, however, and targeting a ban may be administratively costly if enforcement is difficult.

ability that they will be monitored in the future, as outlined by Russell, Harrington, and Vaughan (1986), might also prove quite useful.

Intervening at one stage of the life cycle requires caution to ensure that it does not unintentionally increase exposure at another stage. This difficulty is perhaps best illustrated by examples from the case studies. In the case of formaldehyde, for instance, product labels and standards may reduce potential consumer exposure to the "off-gassing" of formaldehyde contained in various household products. Suppose producers store these products in warehouse inventories for a longer period, so that when the products are eventually delivered to retailers, sufficient off-gassing has already occurred to meet or exceed household standards. An unintended effect of intervention at the end-use level would be that the warehousing may expose warehouse workers to higher levels of off-gassing.

Another example involves incentives to encourage recycling of chlorinated solvents and cadmium. Although incentives may reduce harmful exposure (from disposal) to society in general, emissions during recycling operations may increase harmful exposure to society.

Two additional considerations related to the life-cycle effects of intervention concern the effects of existing regulation. Numerous laws are already in place to regulate various stages of activity related to toxic substances (see figure 1-1), ranging from air and water emissions to transportation and occupational health and safety. An example from the case studies is brominated flame retardants, which are used principally to accommodate fire safety regulations. Intervention to mitigate exposure risk (generally arising from disposal of BFRs) needs to be coordinated with preexisting product safety regulations that induce BFR use; taken together, both sets of regulations need to balance disposal risk with product use.[9]

Generally we note but do not explicitly consider the combined

---

[9]Of relevance to implementation of incentive-based approaches is that, under some interpretations, TSCA appears to allow an integrated approach to regulatory intervention at various life-cycle stages (see figure 1-1, which indicates the purview of TSCA compared with that of other environmental laws). In this case, TSCA could permit the broader approach to regulation. But TSCA also requires the EPA administrator to refer workplace, consumer product, or other stages of exposure to other agencies (for example, the Occupational Safety and Health Administration and the Consumer Product Safety Commission), thus relegating TSCA to a stopgap role (see, e.g., discussion in Doniger [1978] and Shapiro [1990]). The Fifth Circuit Court of Appeals upheld EPA's decision to use TSCA as "a comprehensive statute designed to fight a multi-industry problem" in regulating asbestos (see *Corrosion Proof Fittings* v. *Environmental Protection Agency*, No. 89-4596, slip op. at 23 (5th Cir. 1991).

effect of existing regulations in the case studies. Instead, we assume that the effects of preexisting regulations are separable from the effects of the interventions we outline. Given that a clean slate does not exist for regulating toxic substances, however, the combined effects of regulation might be considered in detail in future extensions of the case studies.[10] Toxic substance regulation must be broad enough in scope to mitigate incentives to adopt potentially more harmful substitutes. The chapters on chlorinated solvents and BFRs emphasize that although substitution away from a toxic substance is a goal of intervention, sometimes the result is substitution of another possibly toxic substance. Approaches such as the posting of a performance bond or some other form of insurance might be considered to guard against the substitution of risky products—or to permit compensation to those harmed (see, e.g., Bohm and Russell [1985]; Costanza and Perrings [1990]; Ippolito [1990]). Such risk sharing by producers may reduce the information burden confronted by regulators of new and existing chemicals.

RATIONALE FOR A CASE STUDY APPROACH

The preceding discussion suggests several reasons why a case study approach is useful in considering applications of incentive-based strategies to toxic substance regulation. One reason is that special tailoring of the design of incentive-based approaches is apparently required to accommodate differences among the toxic substances—there appears to be no "one-size-fits-all" incentive-based regulatory prescription for the whole of toxic substances. The four case studies illustrate the wide variety of circumstances confronting the regulator. Cadmium, chlorinated solvents, and BFRs have exposure properties that call for controlled disposal and, in some cases, recycling. Formaldehyde is notable for its hundreds of end uses where potentially harmful exposure is more or less confined to the user rather than to society as a whole (and where disposal is not a concern). BFRs and

---

[10]In particular, models that outline optimal multiple intervention strategies in the case of municipal solid waste (Menell, 1991a) and multimedia pollution (Oates and Schwab, no date), might be extended to the case of toxic substances. These models generally show that if intervention at one stage (or in one medium) accurately internalizes marginal social damage, then intervention that accurately internalizes marginal social damage at another stage will be welfare enhancing. Specifically, the models indicate that combining incentive-based approaches with performance-based standards or other command-and-control strategies, because the latter are less likely to match marginal damage, may be welfare reducing.

chlorinated solvents are also examples of substances for which substitutes induced by regulatory intervention may be even more worrisome. Incentive-based approaches in these cases, then, call for mechanisms such as deposit-refund to encourage recycling or controlled storage of disposed material (cadmium, chlorinated solvents, BFRs), product labeling and product standard setting (formaldehyde), or performance bonds to encourage safe product substitutes.

Even here, however, differences arise among the approaches recommended for the substances, leading to another reason for a case study methodology. For example, our studies of cadmium and chlorinated solvents—both cases where disposal is a concern—suggest that markets for recycled cadmium and chlorinated solvents are quite different. Cadmium is a by-product of zinc mining and production; thus, its value in a secondary (recycling) market is closely tied to zinc prices. Chlorinated solvents have a high degree of reusability: for instance, closed-loop vapor-degreasing production processes inherently recycle some solvent anyway. Accordingly, different approaches to deposit-refund need to be taken to mitigate environmental or health risk arising from disposal of cadmium or chlorinated solvents. Desirable approaches might encourage additional recycling (for some chlorinated solvents) or controlled disposal incentives that may not seek to induce recycling (cadmium, spent chlorinated solvent sludge, BFRs). These details lead to differentiated policy recommendations that might not otherwise be recognized in the absence of the richness of detail offered by a case study approach.

RELEVANT LITERATURE

A lengthy and still-expanding economics literature argues for the general superiority of incentive-based strategies over command-and-control regulation, primarily because of the relative cost savings expected with incentive approaches. These cost savings arise principally because the latter approaches economize on scarce information about control costs and capitalize on differences in costs among regulatees, give regulatees the incentive to minimize costs of current technology, and provide a basis for environmentally sound and cost-effective technological innovation. Under command and control, these information gaps loom large and can severely limit the cost-effectiveness of regulation. Actual experience with incentive-based approaches offers some, although not unequivocal, evidence of significant cost savings.[11]

Very little of this literature addresses specific issues that arise in the regulation of toxic substances. We focus in this book on specific areas of this literature that are most directly related to characteristics of incentive-based toxic substance regulation, and direct the reader interested in the more general literature on economic incentives to the extensive discussions and bibliographies contained in Bohm and Russell (1985) and Cropper and Oates (1992).[12]

In addition, the large literature on product standards and labeling—not generally associated with the literature on environmental economics—is also relevant to our research.[13] This literature discusses the informational role played by product labeling and standards,

---

[11]Tradable permits to phase down lead in gasoline produced savings of about $200 million per year during a five-year program (see Hahn and Noll [1990] and references therein). The ratio of command-and-control cost to the cost of tradable credits to control air pollution ranges from 1.07 to 22 in studies of provisions of the Clean Air Act summarized by Tietenberg (1985). Cost savings for tradable permits for electric utilities' sulfur dioxide emissions under the Clean Air Act Amendments of 1990 have been predicted to be about $1.5 billion, or 30 percent of an estimated $5 billion in compliance costs under command and control (see ICF Resources, Inc. [1989]).

[12]Roughly speaking, the literature includes four broad topical areas: (1) conceptual models and discussion focusing on efficiency gains of market-based approaches compared with command and control (see, e.g., Dales [1968], Montgomery [1972], Kneese and Schultze [1975], Anderson and coauthors [1977], Baumol and Oates [1988]); (2) extensions of these models addressing complications such as producers' market power (see, e.g., Oates and Strassman [1984]), administrative costs including monitoring and enforcing compliance (see, e.g., Downing and Watson [1974]; Russell, Harrington, and Vaughan [1986]; Russell [1988]), uncertainty in the level of damages or the costs of control (see, e.g., Weitzman [1974], Roberts and Spence [1976]), and general equilibrium results (see Hazilla and Kopp [1990]); (3) research focused on validating the efficiency gains by way of numerical simulation or empirical evaluation of experience with marketlike mechanisms (see, e.g., Tietenberg [1985]; Hahn and Hester [1989]; Oates, Portney, and McGartland [1989]; Hahn and Noll [1990]); and (4) consideration of the political feasibility of marketlike approaches (see Bohm and Russell [1985], Ackerman and Stewart [1988], Hahn and Noll [1990], Breger and coauthors [1991]), and ways to conceptualize equity issues (see, e.g., Burtraw and Toman [1991]). Recent research has begun to consider the integrated effects when intervention takes place at multiple stages of an activity. (Bohm [1981] offered one of the first in-depth analyses of deposit-refund mechanisms; more recently, see Menell [1991a] for a model and discussion of the optimal combination of incentive-based strategies in controlling municipal solid waste; and Oates and Schwab [no date] for approaches to cross-media pollution.)

[13]Tietenberg (1992) addresses product labeling and information in the context of environmental economics. See also Ippolito (1984), Menell (1991b), and Viscusi (1991) for additional perspectives and case studies.

differences in producer responses to voluntary versus mandated product labeling and the setting of product standards; consumers' and workers' responses to this information; and the interplay between regulated product standards and law, particularly product liability. Much of this literature is summarized in the chapters on chlorinated solvents and formaldehyde. For many uses of the substances considered in these chapters, third-party effects are negligible (that is, exposure risk is limited to the end user rather than extending to society as a whole). In the absence of third-party effects, product labeling and standards may play a role in informing consumers of risks that might be associated with a product and in permitting controlled use of or substitution for the product.

The limited literature that contains specific discussion of the application of incentive-based approaches to toxic substances includes Nichols (1982, 1984), Hahn (1988), Industrial Economics, Inc. (1989), U.S. Congress, Congressional Research Service (1989), and U.S. EPA (1991). The Congressional Research Service (CRS) and EPA offer overviews of various examples of the application of these approaches. CRS discusses, among other applications, a deposit-refund system on toxic waste as well as tradable permits for CFC production. The report notes, but does not analyze in depth, some of the advantages and disadvantages of deposit-refund systems. Advantages noted in the report include the self-enforcing property of deposit-refund; disadvantages include employment dislocation effects if recycling reduces the demand for virgin material, administrative costs of the program, and the potential incentive to counterfeit to claim the refund. In the case of CFC permits, the report cites advantages such as the potential for cost savings compared with alternatives such as production quotas, and opportunities for technical innovation.

The EPA report discusses some proposals similar to those we evaluate. These include a deposit-refund system for lead-acid batteries and chlorinated solvents, other incentives to reduce the use of solvents, and marketable permits or taxes for lead production. We generally agree with these approaches and discuss them in depth, although we also offer discussion of some disadvantages and of other alternatives.

In a study of the use of economic incentives to reduce releases of chemicals regulated under the Community Right to Know Act, the Industrial Economics (IEcon) report first categorizes chemicals and then suggests marketlike strategies appropriate to regulating each category. Categories include the number of facilities manufacturing the chemical, market distribution (number of uses) of the chemical, location of the primary end use (geographically concentrated or dispersed), and consumptive versus nonconsumptive use (whether the chemical is an intermediate input, transformed in a reaction, or a final output, as in a solvent application). The report emphasizes, however, that these categories do not fully reflect all chemicals, nor do the decision criteria necessarily lead to the best strategies in all categories. The approach of this book is quite similar in spirit to that report, although we offer more detailed discussion of some of the strategies, discuss some disadvantages, and offer alternatives. Future extensions of our research might build on the approach to chemical groupings taken in the IEcon report so that we might generalize our results for specific chemicals to broader classes of chemicals. Also in the spirit of our research are Nichols's (1982, 1984) case study of the potential cost-effectiveness of emission fees compared with command-and-control regulation in controlling benzene emissions and Hahn's (1988) comparison of economic-based instruments for regulating hazardous waste. Nichols also emphasizes the desirability of targeting intervention, although he focuses on heterogeneous firms rather than heterogeneous outputs.

ORGANIZATION OF THE BOOK

Case studies of chlorinated solvents, formaldehyde, cadmium, and BFRs constitute the next four chapters. As noted earlier, each case study reviews the nature of potential exposure risk, the nature of demand (uses and substitutes for the substance) and supply (essentially, industry structure), and legislation governing the substance. The chapters then discuss how these factors influence the choice of regulatory strategy, offer detailed discussion of the strategy, and indicate why the strategy is preferred to alternative approaches that might be taken. The reference list for each case study cites reference materials on the substances and the relevant economic literature.

The concluding chapter summarizes the studies. It also offers some discussion about the extent to which these case studies might be representative of broader classes of substances and seeks to draw general inferences—regulators' guidelines—about the choice of different incentive-based approaches for different classes. And, finally, we suggest future research directions.

REFERENCES

Ackerman, Bruce A., and Richard B. Stewart. 1988. "Reforming Environmental Law: The Democratic Case for Market Incentives." *Columbia Journal of Environmental Law*, vol. 13, pp. 171–199.

Anderson, Frederick R., Allen V. Kneese, Phillip D. Reed, Russell B. Stevenson, and Serge Taylor. 1977. *Environmental Protection Through Economic Incentives* (Baltimore, MD: Johns Hopkins University Press for Resources for the Future).

Baumol, William J., and Wallace E. Oates. 1988. *The Theory of Environmental Policy*, 2nd ed. (Cambridge, MA: Cambridge University Press).

Bohi, Douglas R., and Dallas Burtraw. 1991. "Utility Investment Behavior and the Emission Trading Market." Discussion Paper ENR91-04 (Washington, DC: Resources for the Future).

Bohm, Peter. 1981. *Deposit-Refund Systems: Theory and Applications to Environmental Conservation and Consumer Policy* (Washington, DC: Johns Hopkins University Press for Resources for the Future).

———, and Clifford F. Russell. 1985. "Comparative Analysis of Alternative Policy Instruments." Pp. 395–460 in Allen Kneese and James Sweeney, eds., *Handbook of Natural Resource and Energy Economics*, vol. 1 (Amsterdam: North-Holland).

Breger, Marshall J., Richard B. Stewart, E. Donald Elliott, and David Hawkins. 1991. "Providing Economic Incentives in Environmental Regulation." *Yale Journal on Regulation*, vol. 8, pp. 463–493.

Burtraw, Dallas, and Michael A. Toman. 1991. "Equity and Effectiveness of Possible CO_2 Treaty Proposals." Draft (Washington, DC: Resources for the Future).

Costanza, Robert, and Charles Perrings. 1990. "A Flexible Assurance Bonding System for Improved Environmental Management." *Ecological Economics*, vol. 2, no. 1 (April), pp. 57–75.

Cropper, Maureen L., and Wallace E. Oates. 1992. "Environmental Economics: A Survey." *Journal of Economic Literature*, vol. 30, pp. 675–740.

Dales, J. H. 1968. *Pollution, Property and Prices: An Essay in Policy-Making and Economics* (Toronto: University of Toronto Press).

Doniger, David D. 1978. *The Law and Policy of Toxic Substances Control (A Case Study of Vinyl Chloride)* (Baltimore, MD: Johns Hopkins University Press for Resources for the Future).

Downing, Paul B., and William D. Watson, Jr. 1974. "The Economics of Enforcing Air Pollution Controls." *Journal of Environmental Economics and Management*, vol. 1, no. 3 (November), pp. 219–236.

Hahn, Robert W. 1988. "An Evaluation of Options for Reducing Hazardous Waste." *Harvard Environmental Law Review*, vol. 12, no. 1, pp. 201–230.

———, and Gordon L. Hester. 1989. "Where Did All the Markets Go? An Analysis of EPA's Emissions Trading Program." *Yale Journal on Regulation*, vol. 6, pp. 109–153.

———, and Roger G. Noll. 1990. "Environmental Markets in the Year 2000." *Journal of Risk and Uncertainty*, vol. 3, no. 4 (December), pp. 351–367.

Hazilla, Michael, and Raymond J. Kopp. 1990. "Social Cost of Environmental Quality Regulations: A General Equilibrium Analysis." *Journal of Political Economy*, vol. 98, no. 4 (August), pp. 853–873.

ICF Resources, Inc. 1989. "Economic Analysis of Title V [sic] (Acid Rain Provisions) of the Administration's Proposed Clean Air Act Amendments" (Washington, D.C., September).

Industrial Economics, Inc. 1989. "Multi-Media Incentives for Reducing Releases of SARA 313 Chemicals." Prepared for the Office of Pollution Prevention, U.S. Environmental Protection Agency, September.

Ippolito, Pauline M. 1984. "Consumer Protection Economics: A Selective Survey." Pp. 1–34 in Pauline M. Ippolito and David T. Scheffman, eds., *Empirical Approaches to Consumer Protection Economics*. Proceedings of a conference sponsored by the Bureau of Economics, Federal Trade Commission, Washington, DC, April 26–27.

———. 1990. "Bonding and Nonbonding Signals of Product Quality." *Journal of Business*, vol. 63, no. 1 (pt. 1), pp. 41–60.

Kneese, Allen V., and Charles L. Schultze. 1975. *Pollution, Prices, and Public Policy* (Washington, DC: The Brookings Institution).

Menell, Peter S. 1991a. "Optimal Multi-Tier Regulation: An Application to Municipal Solid Waste." Draft (University of California, Berkeley, April).

———. 1991b. "The Limitations of Legal Institutions for Addressing Environmental Risks," *Journal of Economic Perspectives*, vol. 5, no. 3 (Summer), pp. 93–114.

Montgomery, W. David. 1972. "Markets in Licenses and Efficient Pollution Control Programs." *Journal of Economic Theory*, vol. 5, no. 3, pp. 395–418.

Nichols, Albert L. 1982. "The Importance of Exposure in Evaluating and Designing Environmental Regulations: A Case Study." *American Economic Review—Papers and Proceedings*, vol. 72, no. 2 (May), pp. 214–219.

———. 1984. *Targeting Economic Incentives for Environmental Protection* (Cambridge, MA: MIT Press).

Oates, Wallace E., and Robert M. Schwab. No date. "Market Incentives for Integrated Environmental Management: The Problem of Cross-Media Pollution." Mimeo (College Park, MD, University of Maryland, Department of Economics, and Bureau of Business and Economic Research).

———, and Diana L. Strassmann. 1984. "Effluent Fees and Market Structure," *Journal of Public Economics*, vol. 24, no. 1 (June), pp. 29–46.

———, Paul R. Portney, and Albert M. McGartland. 1989. "The *Net* Benefits of Incentive-Based Regulation: A Case Study of Environmental Standard Setting." *American Economic Review*, vol. 79, no. 5, pp. 1233–1242.

Roberts, Marc J., and Michael Spence. 1976. "Effluent Charges and Licenses Under Certainty." *Journal of Public Economics*, vol. 5, pp. 193–208.

Russell, Clifford S. 1988. "Economic Incentives in the Management of Hazardous Wastes." *Columbia Journal of Environmental Law*, vol. 13, pp. 257–274.

———, Winston Harrington, and William J. Vaughan. 1986. *Enforcing Pollution Control Laws* (Washington, DC: Resources for the Future).

Schultze, Charles L. 1977. *The Public Use of Private Interest* (Washington, DC: The Brookings Institution).

Seskin, Eugene P., Robert J. Anderson, Jr., and Robert O. Reid. 1983. "An Empirical Analysis of Economic Strategies for Controlling Air Pollution." *Journal of Environmental Economics and Management*, vol. 10 (June), pp. 112–124.

Shapiro, Michael. 1990. "Toxic Substances Policy." Pp. 195–241 in Paul R. Portney, ed., *Public Policies for Environmental Protection* (Washington, DC: Resources for the Future).

Tietenberg, Tom H. 1978. "Spatially Differentiated Air Pollutant Emission Charges: An Economic and Legal Analysis." *Land Economics*, vol. 54 (August), pp. 265–277.

———. 1985. *Emissions Trading: An Exercise in Reforming Pollution Policy* (Washington, DC, Resources for the Future).

———. 1992. *Environmental and Natural Resource Economics* (New York, NY: Harper-Collins).

U.S. Congress, Congressional Research Service. 1989. *Using Incentives for Environmental Protection: An Overview* (Washington, DC, June 2).

U.S. EPA (Environmental Protection Agency). 1991. *Economic Incentives: Options for Environmental Protection*. PM-220 (Washington, DC: U.S. Government Printing Office).

U.S. General Accounting Office. 1984a. *Assessment of New Chemical Regulation Under the Toxic Substances Control Act* (Washington, DC, June 15).

———. 1984b. *EPA's Efforts to Identify and Control Harmful Chemicals in Use* (Washington, DC, June 13).

Viscusi, W. Kip. 1991. "Product and Occupational Liability." *Journal of Economic Perspectives*, vol. 5, no. 3 (Summer), pp. 71–92.

Weitzman, Martin L. 1974. "Prices vs. Quantities." *Review of Economic Studies*, vol. 41, no. 4 (October), pp. 477–491.

2

Chlorinated Solvents

A distinguishing feature of chlorinated solvents (particularly as compared with heavy metals such as cadmium) is their rapid dissipation when exposed to the open atmosphere. This characteristic makes solvents useful in many applications. It also creates the possibility for harmful exposures in some cases while reducing the possibility in others—particularly for solvents whose harmful effects are less likely to affect third parties. The feasibility and cost of reducing dissipation of different solvents in different applications and the presence or absence of third-party effects are important considerations in designing solvent regulation. The extent to which our recommendations for chlorinated solvents can be generalized to other dissipative toxic substances also depends on these cost considerations and third-party effects.

We include four chlorinated solvents in our analysis. They are listed in table 2-1, together with their other chemical names and common acronyms. Perchloroethylene (PERC), trichloroethylene (TCE), and 1,1,1-trichloroethane (TCA) are used primarily in cleaning applications; methylene chloride (METH) is also used for cleaning, but has no single dominant use. The applications for these solvents are discussed in more detail below. Until the early 1990s, CFC-113 was also a major chlorinated solvent, but it is not included in our analysis because its use is being phased out by the year 2000 under the Montreal Protocol on Substances that Deplete the Ozone Layer, and its use has already begun to decline rapidly.

This chapter is organized as follows. After reviewing the sources and uses of each of the chlorinated solvents, we discuss their adverse health and environmental effects and then trace the life cycle of each solvent from production through uses to ultimate disposal. The discussion of product life cycles is organized by specific solvent applications and integrates previously identified harmful effects and exposure paths. It is followed by a brief overview of existing regulations for the solvents.

Table 2-1. The Chlorinated Solvents

Name	Other names	Abbreviation
Perchloroethylene	Tetrachloroethylene	PERC
	Ethelene tetrachloride	
	Tetrachlorethene	
Trichloroethylene		TCE
1,1,1-Trichloroethane	Methyl chloroform	TCA
Methylene chloride	Dichloromethane	METH
	Methelene dichloride	
	Methelene bichloride	

The review of product life cycles and existing regulations sets the stage for identifying the appropriate regulatory incentives to reduce harmful exposures to these substances. We first identify the key features of these substances and the applications that are most relevant for selecting an appropriate regulatory mechanism, and then propose three incentive-based approaches to regulating chlorinated solvents. To complement our proposal, we comment on why other incentive-based approaches are inappropriate, and compare and contrast the mechanisms that we propose with those offered by others (IEI, 1989; U.S. EPA, 1991).

SOURCES AND USES

PERC, TCE, and TCA are produced using the feedstock chemical ethylene dichloride, while METH is produced from methane or methanol (Wolf and Camm, 1987). There are six major U.S. producers, located in Louisiana, Texas, West Virginia, and California. These producers manufacture both the feedstock chemicals and the solvents themselves. Imports of solvents account for less than 10 percent of the domestic market. Solvents are distributed to users through a variety of channels, most of which involve intermediate purchase of the solvent for subsequent sale to a final user. There are many solvent distributors in the United States.

The domestic end uses of virgin chlorinated solvents are listed in table 2-2; percentages do not include sales of reclaimed solvent. Each solvent is discussed briefly below.

Perchloroethylene

Over half of the PERC consumed in the United States is used in dry-cleaning and textile applications. Table 2-2 understates current dry-

Table 2-2. End Uses of Chlorinated Solvents (1989—Estimated)

Chlorinated solvent	End use	Percentage of solvent in end use
Perchloroethylene (PERC)	Dry cleaning and textiles	50
	Chlorofluorocarbon production	30
	Vapor degreasing and metal cleaning	10
	Miscellaneous	10
Trichloroethylene (TCE)	Vapor degreasing and metal cleaning	88
	Polyvinyl chloride manufacture	5
	Solvent and miscellaneous	7
1,1,1-Trichloroethane (TCA)	Metal cleaning	55
	Aerosols	13
	Adhesives	10
	Coatings and inks	10
	Electronics	7
	Miscellaneous	5
Methylene chloride (METH)	Aerosols	25
	Paint remover	25
	Foam blowing	10
	Chemical processing	10
	Metal cleaning	10
	Electronics	10
	Miscellaneous	10

Sources: Mannsville Chemical Products Corporation, *Chemical Products Synopsis: Methylene Chloride* (September 1988), *Chemical Products Synopsis: Perchloroethylene* (February 1989), *Chemical Products Synopsis: Trichloroethylene* (February 1989), *Chemical Products Synopsis: 1,1,1-Trichloroethane* (October 1990) (Asbury Park, NJ: Mannsville Chemical Products Corporation).

cleaning use because declining production of CFC-113 has reduced demand for PERC as a feedstock but has increased the demand for PERC in dry-cleaning applications to replace CFC-113. PERC is also used in vapor degreasing and other metal-cleaning applications. About 10 percent of the total quantity of domestically produced PERC was exported in 1988.

Trichloroethylene

This substance is used primarily in metal cleaning and vapor degreasing. TCE is also an input into the production of polyvinyl chloride (PVC) and nonflammable adhesive formulas. Approximately one-third of total domestic production was exported in 1988.

1,1,1-Trichloroethane

State regulation of TCE emissions from metal-cleaning operations has led to increased substitution of TCA for TCE in metal cleaning, thus contributing to TCA's large share of this market (Chestnutt, 1988). TCA is also used in aerosol production as well as in electronics, primarily for the cleaning of printed circuit boards. Approximately 10 percent of domestically produced TCE solvent was exported in 1988.

Methylene Chloride

METH has a diversity of uses. Emissions from applications, such as aerosols and home paint stripping, dissipate totally, whereas emissions from such applications as industrial and commercial paint removal and electronics applications are dependent upon the technology employed. Exports in 1988 equaled about one-fourth of domestic production, a substantially higher percentage than past trends of around 10 to 15 percent.

HEALTH AND ENVIRONMENTAL EFFECTS

Health and environmental effects of chlorinated solvents, summarized in table 2-3, differ by product. All of the solvents are harmful to skin and eyes and may produce adverse effects on the central nervous system (CNS), although somewhat higher exposures are required for CNS damage from TCA and METH. TCE contributes to the formation of tropospheric ozone and smog. TCA contributes to

Table 2-3. Health and Environmental Effects of Solvents

Solvent[a]	Smog precursor	Central nervous system damage	Skin/eye irritant	Ozone layer depletion	Carcinogenic
PERC	No	Yes	Yes	No	Possibly
TCE	Yes	Yes	Yes	No	Possibly
TCA	No	Yes[b]	Yes	Yes	Testing
METH	No	Yes[b]	Yes	No	Probably

[a]PERC = perchloroethylene; TCE = trichloroethylene; TCA = 1,1,1-trichloroethane; METH = methylene chloride.

[b]High-level exposures are necessary for harmful effects.

Source: Halogenated Solvent Industry Alliance, *Chlorinated Solvents in the Environment* (Washington, DC, 1986).

the depletion of the stratospheric ozone layer, although at a fraction of the potency of CFCs. PERC and TCE have been classified as possible carcinogens, METH is a probable carcinogen, and TCA testing continues.

LIFE CYCLE IN VARIOUS APPLICATIONS

The life cycle of a typical chlorinated solvent is outlined in table 2-4. This table traces each solvent from production through distribution and use to disposal and indicates the potential exposure problems that arise at each stage for the different applications. In general, harmful exposure and release problems do not arise until the solvent is used in a cleaning or another potentially more dissipative application. Depending on the application, releases can be harmful to the user, harmful to the environment and third parties, or both. Exposure problems also may occur at the disposal stage, largely as a result of improper disposal of spent solvent or sludge from a cleaning process that contains a high concentration of solvent.

Table 2-4. Life Cycle and Exposure

Stage of production or use	Solvent[a]	Exposure
Feedstock	All	Workplace
Production	All	Workplace
Distribution	All	Ambient air, water by way of accident and spills
Primary use		
Dry cleaning	PERC	Workplace (skin and air) Ambient air
Metal cleaning	TCE	Workplace (skin and air) Ambient air
Metal cleaning	TCA	Workplace (skin and air) Ambient air Stratospheric ozone
Paint stripping and aerosols	METH	Worker (skin and air) Consumer (skin and air) Ambient air
Recycling	All	Workplace (air) Ambient air
Disposal	All	Groundwater Ambient air

[a]PERC = perchloroethylene; TCE = trichloroethylene; TCA = 1,1,1-trichloroethane; METH = methylene chloride.

The following paragraphs briefly discuss each of the primary applications, harmful exposure problems, and potential substitutes for these solvents. The two primary sets of applications covered here are (1) commercial and industrial cleaning—which includes dry cleaning, cold metal cleaning and vapor degreasing, paint stripping during original equipment manufacturing (OEM), maintenance paint stripping, and cleaning of printed circuit boards; and (2) dissipative uses by consumers—which include aerosols and commercial as well as do-it-yourself (consumer) paint stripping. The main distinction between these two categories of uses (with some exceptions) is recoverability. Applications in the first category lend themselves to partial recovery of spent solvent for recycling, whereas applications in the second category generally do not. The potential recoverability of a solvent in a particular application is a salient factor in determining the appropriate incentive mechanism, as discussed below.

Dry Cleaning

PERC is transported from the manufacturer to more than 80 percent of the approximately 30,000 dry-cleaning establishments across the country by a large number of dispersed retail and distribution companies. Dry cleaners use solvent instead of soap and water to clean clothes. Dry-cleaning activity leads to two types of releases: vapor and spent solvent.[1] The amount of solvent lost depends on the technology used by the dry cleaner. Releases are greater in the older system, which required that clothes be transferred from one machine to another between the wash and dry stages, than in the dry-to-dry or closed system. The transfer process results in releases to the workplace and, possibly, to outdoor air. However, carbon filters can be used to capture PERC emissions before they escape to the air outside, and the solvent subsequently can be reclaimed through a process known as steam stripping (HSIA, 1989). In the closed system, emissions are reduced because the vaporized solvent is constantly being condensed and reused. The classification of PERC as an ozone precursor and the subsequent introduction of emissions limitations in state implementation plans mandated by the Clean Air Act Amendments of 1970 have led to more efficient use of virgin PERC and a reduction in the volume of PERC emissions from dry cleaning (Wolf and Myers, 1987).

---

[1]Some solvent is released into the water, but the amount is a negligible fraction of all solvent used in dry cleaning (Wolf and Myers, 1987).

The quantity of spent solvent that remains for ultimate disposal can be reduced by internal distillation processes that essentially boil off the solvent content of the solvent–contaminant mixture and condense it for future use (HSIA, 1989). Such solvent recycling is practiced in-house by a number of dry cleaners as well as externally by waste handling/recycling firms. Since at peak efficiency, recycling only recovers 90 to 95 percent of the solvent content of the sludge, the "still bottoms" that remain after distillation must be disposed of in an environmentally safe manner. Under the Resource Conservation and Recovery Act (RCRA), still-bottom disposal is limited to hazardous waste incinerators or, after appropriate treatment to reduce solvent concentrations to allowed levels, to land-based RCRA facilities. (See table 2-5.) Still bottoms also can be burned as fuel at cement kilns (Morton, no date).

Currently, there are no feasible chemical or process substitutes for PERC in dry cleaning. Technical substitutes include CFC-113 and various flammable carbon-based solvents. The former is being phased out under the Montreal Protocol and the latter are illegal under the National Fire Protection Codes. Thus, the only real alternative is the substitution of washable fabrics for nonwashable ones.

Cold Cleaning of Metals

All of the solvents are used in metal cleaning; however, the primary solvent for this application is TCE. Thousands of machinists, auto repair shops, metal furniture manufacturers, and metal parts fabricators across the country use these solvents to remove dirt and grease from metal parts (Pekelney, 1991). Metal items are cleaned by subjecting them to vapors at low temperatures or by dipping them in a tank of solvent. Either a batch process in which a number of items are cleaned at one time or a continuous process in which items pass through the machine on a belt can be used. Worker exposures to very high concentrations tend to be greater with a batch process, although overall releases tend to be higher with a continuous process. Filters similar to those used by dry cleaners are used to absorb emissions before they leave the plant. The cleaning process also produces a sludge that may be disposed of or distilled to reclaim used solvent, leaving still bottoms for ultimate disposal.

The substitutes for cold metal cleaning include aqueous and semiaqueous processes. These water-based methods have adverse effects of their own, including the potential of rusting, the creation of residual films, the creation of water pollution, and a longer drying

Table 2-5. Statutes and Regulations Pertaining to Chlorinated Solvents: PERC, TCA, METH, and TCE

Responsible agency	Statute, regulation, or agreement	Uses or sources of exposure covered
FEDERAL		
Consumer Product Safety Commission	Federal Hazardous Substances Act	Brings action against products containing METH that are not labeled as possible carcinogens.
Food and Drug Administration	Delaney Clause of the Federal Food, Drug, and Cosmetic Act	Defines METH as decaffeinating agent rather than as food additive to avoid regulation under Delaney Clause.
Occupational Safety and Health Administration (OSHA)	Occupational Safety and Health Act	Limits exposure in workplace: standards are 25 parts per million (ppm) PERC, 500 ppm METH, 350 ppm TCA, and 50 ppm TCE.
U.S. Department of the Treasury	Excise Tax of 1991	Taxes imports of PERC and TCE; TCA taxed as an ozone-depleting chemical.
U.S. Department of Transportation	Hazardous Materials Transportation Act	Regulates transportation of all four substances as chemical or hazardous waste.
U.S. Environmental Protection Agency (EPA)	Clean Air Act	Mandates setting of national emissions standards for all four chlorinated solvents, as all are defined as hazardous air pollutants (section 112).
EPA	Clean Water Act	Requires EPA to set standards for all four solvents. Half-lives for degradation of solvents in water range from 1.2 to 2.8 years.
EPA	Comprehensive Environmental Response, Compensation, and Liability Act	Lists all four chlorinated solvents as hazardous under section 101(14).
EPA	Federal Insecticide, Fungicide, and Rodenticide Act	Requires data on the use of METH in pesticides.
EPA	Resource Conservation and Recovery Act	Bans land disposal of untreated wastes; exempts waste treated to meet EPA standards, which are (mg/liter) TCE 0.062, PERC 0.079, TCA 1.05, METH 0.2.

(continued)

25

Table 2-5 (*continued*)

Responsible agency	Statute, regulation, or agreement	Uses or sources of exposure covered
EPA	Safe Drinking Water Act	Sets standards of 5 µg/liter of drinking water for TCE and 0.2 mg/liter for TCA.
EPA	Toxic Substances Control Act	Lists METH under section 4(f), which designates it for special review.
STATE		
	California Proposition 65	Requires businesses in California to warn consumers and workers of dangers of exposure to TCE, PERC, and METH.
	State regulations	Regulate TCE and PERC as ozone precursors under many state implementation plans.
INTERNATIONAL		
	Montreal Protocol Amendments	Phase out production of TCA by 2005.

Note: PERC = perchloroethylene, TCA = 1,1,1-trichloroethane, METH = methylene chloride, and TCE = trichloroethylene.

Sources: Kathleen Wolf and Frank Camm, *Policies for Chlorinated Solvent Waste—An Exploratory Application of a Model of Chemical Lifecycles and Interactions* (Santa Monica, CA: RAND Corp., June 1987); Halogenated Solvents Alliance, *Chlorinated Solvents in the Environment* (Washington, DC: February 1989); Mannsville Chemical Products Corporation, *Chemical Products Synopsis: Methylene Chloride* (September 1988), *Chemical Products Synopsis: Perchloroethylene* (February 1989), *Chemical Products Synopsis: Trichloroethylene* (February 1989), *Chemical Products Synopsis: 1,1,1-Trichloroethane* (October 1990) (Asbury Park, NJ: Mannsville Chemical Products Corporation); Internal Revenue Service, *Excise Taxes for 1991* (Washington, DC: U.S. Department of the Treasury, 1991).

time (HSIA, 1989; Pekelney, 1991). Hydrocarbon solvents are also a possible substitute, but their use is limited by local fire codes.

Vapor Degreasing

Both TCE and PERC are used for vapor degreasing. This process involves heating solvent and running the hot vapors over the grease-laden materials (glass, metal, or other nonporous materials), which are introduced to the process at room temperature. The solvent condenses to a liquid when it comes into contact with the cool material, and gravity causes the condensed-solvent–laden grease to drip into a container (ASTM, 1989). Generally, this process is more closed than cold cleaning, limiting air emissions. Nevertheless, some emissions escape when materials are introduced into the process, resulting in worker exposure and releases into the atmosphere. In addition, the process creates grease-laden spent solvent to be distilled and the still bottoms for disposal. More than half of all degreasers use on-site recycling to reclaim spent solvent.[2] The substitutes for this type of cleaning are similar to the substitutes for cold metal cleaning and, therefore, suffer from similar drawbacks. However, many electrical equipment manufacturers, anxious to avoid the high cost of complying with the Clean Air Act and the potential lawsuits associated with workplace exposure, are switching to aqueous or semiaqueous cleaning methods that do not make use of chlorinated solvents.[3]

Electronics—Circuit Board Defluxing and Cleaning

TCA and METH are used in the electronic industry by circuit board manufacturers to remove flux introduced in the soldering process as a surface-tension-reducing measure and to remove photographic-resistant chemicals (Pekelney, 1991). The exposure paths for workers employed in this process are similar to those in metal cleaning and dry cleaning. Workers are exposed when products are removed from, or placed into, the defluxing machine. Atmospheric releases beyond the plant also occur, but these can be reduced by use of filters. Recycling can be used to reduce the volume of the spent solvent waste.

Process and product substitutes that do not require defluxing can replace this application of solvents. Aqueous cleaning is a possibility, although it leaves residues that affect the performance of circuit boards. Product substitutes include printing circuits directly onto a product, thus eliminating the need for boards and soldering, or reducing the reliability standards for circuit boards in less precise applications, such as consumer electronics (Wolf, 1990a).

Paint Removal—OEM and Maintenance

The sole solvent used in paint removal applications is METH. The OEM application involves removing paint from painting equipment

---

[2]Personal communication from Steve Rissoto, Halogenated Solvent Industry Alliance, Washington, D.C., May 14, 1991.

[3]Personal communication from Fred Anderson, consultant to National Electrical Manufacturers Association, Washington, D.C., July 18, 1991.

and from sprayed areas where paint has been applied to large items such as new cars (Wolf, 1990b). Also included here is the removal of paint from a defective part before it is reworked into a usable part. The maintenance application involves the removal of paint from planes, ships, tanks, and even automobiles as a part of a machine maintenance program. A variety of methods can be used to apply the paint remover: spray, brush application, or dipping (for small parts). Worker exposure routes are similar to those for other cleaning applications. The harm from atmospheric releases is relatively small, since this chemical is neither a tropospheric ozone precursor nor a stratospheric ozone depleter and it dissipates quickly, thus minimizing the chance of third-party exposures large enough to pose a cancer risk to the population at large.

The disposal problems associated with this process relate to both the removed paint and the solvent. If a dipping process is used, reclamation possibilities exist. Some solvent also may be reclaimed from carbon filters used to reduce air emissions, but the tendency of METH to produce corrosive reactions in the presence of steam greatly limits the use of standard steam stripping technology for this process (HSIA, 1989).

Chemical substitutes for paint removal offer some promise for OEM where, generally, paint is not fully cured; however, they are virtually nonexistent in the maintenance market. Process substitutes include abrasive methods and cryogenics, but these are difficult to apply to large surfaces and may damage the underlying material.

Paint Removal—Consumer Market

The solvent used in this market, which includes commercial furniture strippers and do-it-yourself paint removal, is also METH. Commercial paint strippers dip items in a stripping tank. The process results in 80 percent vaporization of the solvent with 20 percent left for disposal (Wolf, 1990b). The rate of vaporization can be reduced somewhat by using a closed system and reclaiming vapors captured in filters. Spent solvent in the tank can be made reusable by filtering out contaminants or recycling by use of distillation methods. Illegal disposal of spent solvent does not pose a serious threat to groundwater, since METH biodegrades quickly (HSIA, 1989). Substitutes for this form of paint removal include cryogenic methods, which are expensive and may damage furniture (Wolf, 1990b).

Do-it-yourself paint removal results in significant exposures to the user but, generally, few third-party effects. Virtually 100 percent of the solvent used in this application is released into the atmosphere (Wolf, 1990b). Harmful effects due to consumer exposures can be reduced through the use of adequate ventilation. Disposal of small quantities of the solvent in a traditional landfill does not pose a serious threat to groundwater since the solvent biodegrades quickly when introduced into groundwater (HSIA, 1989).

Aerosols

Both TCA and METH are used in aerosols, the primary applications of which are spray paints and lubricants (Pekelney, 1991). Users experience some exposure that might be of some concern, particularly for METH, which has been classified a probable carcinogen. Atmospheric releases beyond the immediate user are more of a problem from TCA, an ozone-depleting substance, than from METH. Final disposal could also be a problem for incompletely used TCA-containing aerosols. TCA does not break down in groundwater as quickly as METH and, thus, disposal of residual TCA-containing aerosols in landfills could present a problem.

EXISTING STATUTES AND REGULATIONS

The current statutes and regulations governing environmental releases and ambient air standards for chlorinated solvents are listed in table 2-5. A major exposure route for solvents used in cleaning applications is vapors in the workplace. The Occupational Safety and Health Administration (OSHA) has set standards for ambient air concentrations for all four solvents in an effort to reduce worker exposures and their harmful effects.[4] All four substances are classified as hazardous air pollutants under the Clean Air Act, and both TCE and PERC are regulated as ozone precursors with emissions limited under most state implementation plans. (For some states these emissions are limited only in ozone nonattainment areas—that is, regions of the country not in compliance with the National Ambient Air Quality Standards for ozone.) The Clean Water Act and the Safe Drinking Water Act limit emissions and concentrations in water, respectively.

Both the transportation and final disposal of spent solvent or

---

[4]Given the large number of establishments where solvents are used, adherence to these standards may be difficult to monitor and violations difficult to detect.

solvent-laden sludge are also regulated. The Hazardous Materials Transportation Act governs the transport of spent and virgin solvents. The Resource Conservation and Recovery Act prohibits land disposal of any of the four solvents unless they have been treated using the Best Demonstrated Achievable Technology to result in the concentration levels indicated in table 2-5. Treated spent solvent may not be sent to a municipal landfill; its disposal must be at a RCRA facility. Untreated waste may be incinerated at a licensed hazardous waste incinerator or in a cement kiln. The imposition of the land disposal ban has greatly increased the cost of disposing of still bottoms from solvent recycling, thereby reducing the level of in-house recycling activity (Wolf and Camm, 1987) and increasing the level of external recycling (Pekelney, 1990).

SOLVENT ATTRIBUTES RELEVANT TO ECONOMIC INCENTIVE DESIGN

The goal of regulating chlorinated solvents is to reduce potentially harmful exposures. The preceding analysis of the life cycle of solvents in their different major applications indicates that potentially harmful exposures are associated with product use and, in many applications, with product disposal. The exposure problems created during product use include both harmful exposure to consumers or workers actually using the solvent or solvent-containing product, and harmful exposure to third parties. These two features—when exposure takes place and who is potentially at risk from exposure—are key to selecting an appropriate regulatory mechanism.

The structure of the chlorinated solvents industry with its few producers, many more distributors, and very large number of end users also has implications for the selection of a regulatory mechanism. Aiming a regulatory policy for reducing third-party exposure toward solvent producers may be administratively more efficient than directing the policy toward solvent users, whose behavior determines the level of emissions. The reusability of solvent condensate and recycled spent liquid solvent suggests that the allocative efficiency losses from indirect targeting of the regulation may be fairly small. Reusability implies that properly specified incentive mechanisms directed at solvents as inputs, such as a tax on solvent sales, can closely approximate an incentive mechanism directed at air emissions or, to a lesser extent, at disposal activity. By increasing the price of primary solvent, such a tax increases the solvent user's incentive to contain emissions and recycle spent liquid solvent.

Another important feature of this industry that should guide the design of new environmental regulations is the uncertainty about the benefits of additional reductions in exposure, given existing regulations. If existing regulations are effective in reducing workplace exposures, limiting atmospheric releases, and preventing the harmful effects of disposal in municipal landfills, are new regulations necessary? The answer to this question may be that the costs of existing command-and-control regulations, including the high costs of enforcing compliance by a large number of widely dispersed solvent users and disposers, may be reduced by the introduction of an incentive-based regulatory mechanism. For example, given the high costs of compliance with existing disposal regulation and the associated disincentives for compliance, a new regulatory regime may be in order if it can both increase regulatory compliance and decrease enforcement costs.

PROPOSED INCENTIVE MECHANISMS

We focus on three methods of solvent regulation to address the three main sources of potentially harmful exposure: emissions into the atmosphere, air emissions in the proximity of users, and releases associated with improper disposal. In each case the purpose of the regulation is to reduce the extent of potentially harmful exposures in a cost-effective manner. As indicated above, the potential harm associated with these three sources of exposure varies across chlorinated solvents, and, for a particular solvent, across solvent applications. Therefore, different regulatory interventions are appropriate for different solvents in their different applications. Table 2-6 presents regulatory options for each of the three exposure points, along with reasons why each is appropriate or inappropriate.

In the following paragraphs, we present the proposed mechanisms in greater detail and discuss the incentive effects of these mechanisms as well as other important features of the proposed regulations. We also compare and contrast the selected mechanisms with alternatives that are less appropriate for regulating chlorinated solvents and with mechanisms proposed elsewhere.

Excise Tax on Sales of Solvents

One approach is to levy a virgin product tax on producers and importers of virgin chlorinated solvents. Ideally, the tax would be

Table 2-6. Regulatory Options for Chlorinated Solvents

Exposure point	Regulatory option	Verdict	Comments
Product manufacture and use			
Air emissions and third parties (noncontainable applications)	Labeling	N.A.	
	Emission tax	No	Difficult to monitor emissions.
	Tradable emission permits	No	Too many users.
	Product tax	Yes	Sales of product easy to monitor.
	Deposit-refund	N.A.	
	Command and control	No	Might be advised if administrative cost of tax is very high.
Air emissions and third parties (potentially containable applications)	Labeling	N.A.	
	Emission tax	No	Too expensive to monitor emissions with large number of dispersed users.
	Tradable emission permits	No	High monitoring cost.
	Product tax	Yes	Product purchase easy to monitor—few producers/sellers relative to users. Also, taxing product raises cost of air emissions since captured emissions can be substituted for new solvent.
	Deposit-refund	N.A.	
	Command and control (emission standard)	No	Might be advised if administrative cost of tax is very high.

Product user	Labeling	Yes	No externality is associated with use, and goal is better-informed user.
	Emission tax	No	No externality.
	Tradable emission permits	No	No externality.
	Product tax	No	No externality.
	Deposit-refund	N.A.	
	Command and control (content limit)	Maybe	If consumer is unable to understand labels, or to respond to labels. (Politically more attractive than labels.)
Disposal	Labeling	No	No incentive effect.
	Emission/disposal tax	No	Provides disincentive for proper disposal and may lead to dumping.
	Tradable emission permits	No	Same incentive to dump as tax; may have high administrative/enforcement cost.
	Tradable recycled-content permit	No	Virtually impossible to set target percentage of recycled content.
	Deposit-refund	Yes	Provides incentive for proper disposal.
	Command and control (disposal requirements)	No	May provide a disincentive for proper disposal if cost is high enough and illegal dumping has relatively low expected cost.

Note: N.A. = not applicable.

33

set at a level equal to the marginal environmental and health damages associated with solvent emissions. The size of the tax would vary by solvent to reflect the relative differences in damages associated with emissions of different solvents. Levying different tax rates on different solvents might lead consumers to substitute one solvent for another, but presumably this substitution would be away from a higher-taxed and thus more-damaging solvent toward the lesser-taxed and thus less-damaging one.

To the extent possible, the tax might also vary by solvent application to reflect the different marginal emission damages associated with different uses. This type of tailoring may be difficult, since a consumer could purchase a low-taxed solvent for a stated use and then actually use it in a riskier application. If the costs of policing such substitution in use are high, an identical tax schedule across the relevant set of applications may be desirable.[5] The tax would not be levied on sales of recycled solvent, because the damages associated with releases of any recycled solvent would presumably be captured in the tax levied initially on the virgin solvent from which the recycled solvent is derived.

The incentive effects of this tax will depend on the application of the solvent and the potential for containing the resulting air emissions. The tax will increase the cost of solvent use in all applications and, thus, will encourage users to search for less harmful substitutes that, presumably, will be subjected to less regulatory scrutiny. In applications such as paint stripping by consumers and aerosol products, where the solvent is virtually 100 percent dissipated in use, there is no opportunity for recycling. Without regulation, society suffers a social welfare loss due to the externality associated with using these solvent-containing products. The excess consumption associated with failure to internalize this externality is illustrated in figure 2-1.

A product tax equal to the marginal damage of air emissions will reduce quantity demanded to the point where marginal private benefit equals marginal social cost. The welfare gains from instituting this tax are illustrated in figure 2-1. In this graph, the marginal cost of producing a solvent is assumed to be constant at MPC. Given the demand curve represented by D, the quantity of solvent demanded is $\hat{Q}_s$. The marginal social cost of producing solvent is given by MSC. Since at $\hat{Q}_s$, MSC lies above the demand curve (depicting marginal private benefits of consumption), there is a welfare loss equal to the

---

[5]We plan to consider more formally the efficiency implications of tailoring production taxes to accommodate heterogeneous uses in future research.

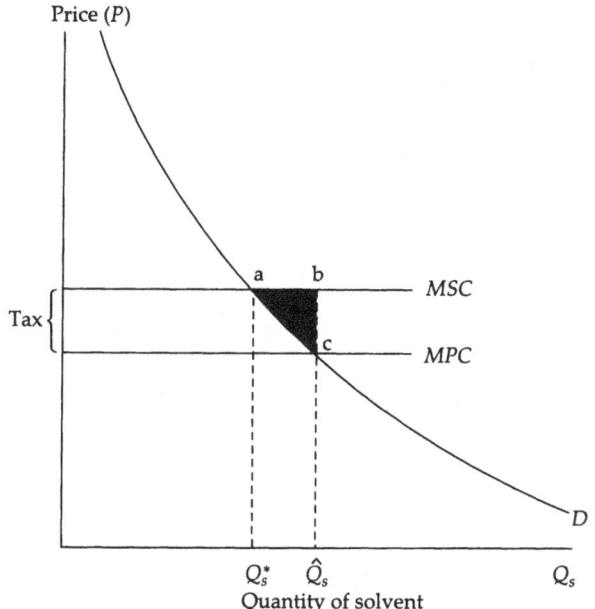

Figure 2-1. Effects of a solvent tax on dissipative uses.

area of the triangle *abc*. When a tax equal to the marginal damage cost ($MSC - MPC$) is added to the price, the quantity demanded is reduced to Q_s^*. At this output level, marginal social cost and marginal private benefits are equal and social welfare is maximized.

The effects of the tax are somewhat more complex in the less dissipative applications. In most industrial and commercial cleaning applications of solvents (including dry cleaning, metal cleaning, vapor degreasing, and cleaning of electronics equipment), emissions are at least partially containable. The emission control technologies either result in a longer useful life for a given-size batch of solvent in its current application or require a separate intermediate step in which solvent contained in filters and other devices is reclaimed for future reuse. Given that emission reduction increases the productive life of a given quantity of solvents, increasing the price of virgin solvents will increase the user's desire to reduce emissions, reclaim captured and contaminated (spent) solvent, and reduce reliance on virgin solvent.

The efficiency gains from a damage-based tax on sales of virgin chlorinated solvent and its effects on the level of solvent recycling and emission control activity can be illustrated with a simple graphic model. The graph in figure 2-2 is derived from a two-period model

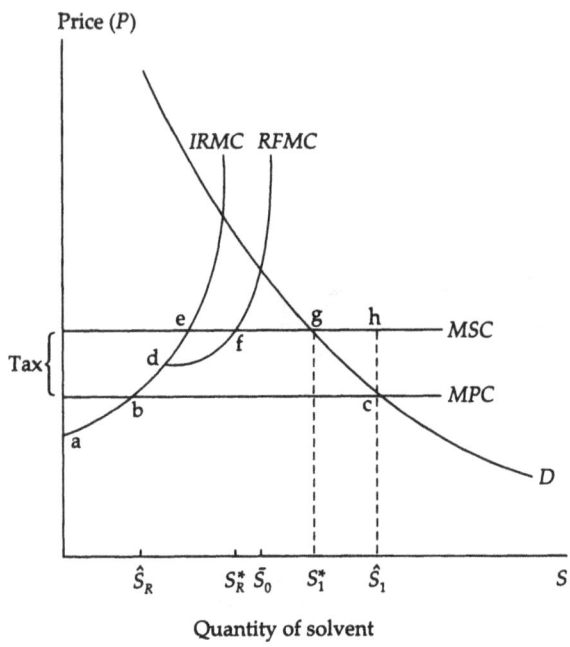

Figure 2-2. **Effects of a product tax with partially containable uses.**

of the market for a chlorinated solvent to be used in a particular cleaning application. The graph shows the state of the market in the second period. The amount of spent solvent left from the previous period and available for potential recycling is labeled $\bar{S}_0$.

The constant marginal cost of producing virgin solvent is illustrated by the line labeled MPC. The marginal cost of recycling solvent internally is given by the curve $IRMC$ (internal recycling marginal cost) and the marginal cost of recycling solvent at a full-time recycling facility is $RFMC$ (recycling facility marginal cost). Note that these two supply curves become asymptotically vertical at a point where each respective technology is unable to extract any more useful solvent from the existing stock of spent solvent, $\bar{S}_0$. The superior efficiency of external recycling is evidenced in the graph by the flatter initial slope of the $RFMC$ curve and the higher full extraction capability. If the price of virgin solvent equals the constant private marginal cost, then the effective supply curve for solvent is given by the curve abc. Total quantity of solvent demand is given by $\hat{S}_1$ with $\hat{S}_R$ being recycled solvent and $\hat{S}_1 - \hat{S}_R$ being virgin solvent.[6]

---

[6]This graph clearly shows that competition from recycled solvents disciplines the

The marginal social cost of solvent is assumed to be greater than the marginal private cost because of the environmental and health damages associated with solvent releases into the environment. If the marginal social cost is given by the curve labeled MSC in figure 2-2, then failure to incorporate this externality into the costs faced by the solvent user lead to a social welfare loss equal to the area of triangle ghc.

If sales of virgin solvent are taxed to equate consumer price of virgin solvents to marginal social cost, the social welfare loss is eliminated. Under such a tax, a positive quantity of solvent is supplied by the professional solvent recycler and the effective supply curve becomes $aefh$.[7] The total quantity of solvent demanded falls to S_1^*, the quantity of recycled solvent supplied rises to S_R^*, and the quantity of virgin solvent sold falls to $S_1^* - S_R^*$. The addition of a product tax will lead to increased recycling of the existing stock of spent solvent unless the current cost structure is such that the MPC curve intersects the $IRMC$ and $RFMC$ curves beyond the point where these supply curves reach their physical limit.

A solvent tax may also create an incentive to reduce emissions of solvent during cleaning applications. Figure 2-3(B) shows the technical and cost tradeoff between inputs of solvent (S) and emissions reduction capital (ERK) in the production of cleaning services. The curve labeled I^* represents an isoquant for cleaning services, the slope of which is the marginal rate of technical substitution between ERK and S.[8] The budget line absent a tax on solvent is given by ac.

When a tax is added to sales of virgin solvent, the effective price of solvent increases and the budget line rotates in toward the vertical axis to become ae. The tax will lead the firm to use more ERK relative to S. If output of cleaning services is held fixed at I^*, the tangency moves from $\hat{T}$ (input levels $\hat{E}$ and $\hat{S}$) to T^* (input level E^* and S^*).

---

market for virgin solvent. The role of recyclers in providing competition for virgin solvent producers has been the subject of many antitrust cases, beginning in 1945 with *United States* v. *Aluminum Corporation of America*, 148 F.2d 416 (2d Cir. 1945). In that case, the court failed to recognize that recycled aluminum is a substitute for primary aluminum and therefore should be included in the market definition. For a formal analysis of the impacts of recycling on the pricing behavior of primary-goods monopolists, see Gaskins (1974), Martin (1982), and Tirole (1988).

[7]If recycled solvent is a perfect substitute for virgin solvent, commercial recyclers will charge the virgin solvent market price for their product. With a virgin product tax, the solvent user will employ internal recycling methods until the marginal cost of recycling internally equals MSC. Assuming solvent users can send the rest of their spent solvent to a professional recycler, they will purchase an additional ef units of professionally recycled solvent at a price of MSC.

[8]Along these isoquants, the quantity of the other inputs such as capital for cleaning purposes or labor are assumed to be adjusted optimally.

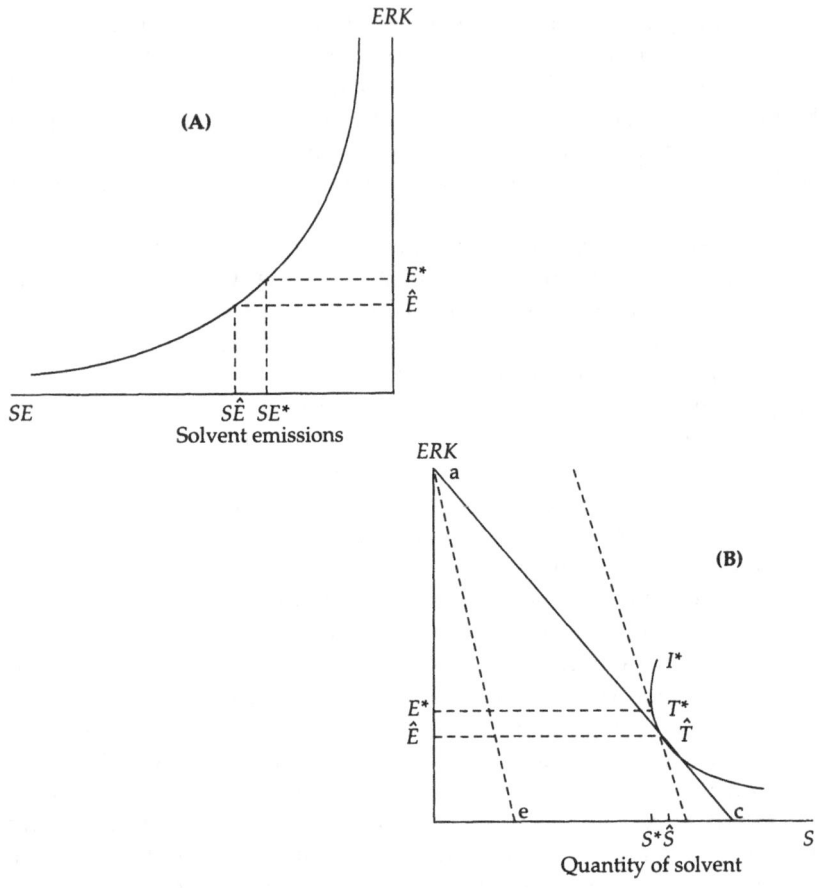

Figure 2-3. Effects of a product tax on solvent emissions.

The graph in figure 2-3(A) illustrates the emission reduction function and shows how increasing the amount of emission reduction capital leads to a decrease in the level of solvent emissions from $S\hat{E}$ to SE^*.

In addition to its desirable incentive effects, the proposed product tax has low administrative costs because of the small number of producers and solvent importers who would be subject to the tax.[9]

---

[9]In an earlier study of incentive-based regulation, IEI (1989) proposed a product-based tax on sales of solvent to be imposed at the point of final sale instead of on the manufacturer. Although such a decentralized tax is more easily tailored to match regional differences in damage costs, it would be more expensive to administer than the tax proposed here. If taxes were allowed to differ across regions, there might be

Two other approaches that generate incentives for emission reduction—emission taxes and permit trading—were rejected because of their high administrative costs for achieving the same level of benefits associated with the product tax.

A tax on actual emissions instead of a tax on sales of the product could require costly emission monitoring to ensure compliance. The monitoring requirements would be reduced somewhat if the tax were levied on estimated emissions, where the latter depended on the installation of control equipment and other emission reduction strategies as well as the level of solvent use.[10] Basing the tax on self-reported emissions will create an incentive for underreporting of emissions. With a tax on estimated emissions, the firm will have an incentive to underreport output levels or input use to avoid the emission tax. This incentive could be reduced by instituting random audits and fines for misreporting or by requiring firms to install continuous emission monitoring (CEM) equipment at all sites. Both options, particularly CEM equipment, which would require regular meter readings by local environmental officials, would prove very expensive. Given these enforcement difficulties for an emission tax, a product tax with its strong incentives for emission reduction and relatively low administrative costs is a more attractive option. A product tax also has the ancillary benefit of increasing the direct cost of solvent spills and releases within the workplace that may not be present with an emission tax. These benefits may reduce noncompliance with existing OSHA regulations at essentially no additional cost and reduce the need for OSHA compliance monitoring.

An emission permit trading scheme that allowed chlorinated solvent users to bid for the right to emit solvent vapors into the atmosphere would also require some mechanism for monitoring actual or estimated solvent emissions and ensuring that permitholders'

---

an incentive, depending on the size of the differences and transportation costs, to purchase solvent in a region with a lower tax rate than the region where it is to be used. This possibility could limit the regulator's ability to target a product tax to reduce emissions in a particular region. The U.S. EPA (1991) proposed a product tax on distributors of all products emitting volatile organic compounds (VOCs), with the tax related to the VOC content of the products. This tax would apply to all solvents classified as ozone precursors (one of the major harmful features of TCE), but would not address other toxic effects or the substitutability of other solvents for TCE. Chestnutt (1988) has pointed out how Clean Air Act restrictions on TCE have led to the substitution of TCA, an ozone-depleting substance.

[10]The efficacy of estimating emissions on the basis of investment in control technologies will depend on the relative size of fixed and variable costs for operating the emission control equipment. A large variable cost may provide a disincentive for abatement.

emission levels actually match their permitted levels.[11] Given the problems with monitoring emissions, a permit trading scheme would also prove more administratively burdensome than a product tax.[12] Nonetheless, permit trading schemes have been adopted to implement VOC emission reductions under the Clean Air Act in some nonattainment areas, including the South Coast Air Quality Management District (SCAQMD) in California.[13]

Labeling to Inform Consumers

For some solvent applications, such as METH in consumer paint strippers, the potential third-party effects may be small, and therefore labeling may be the most appropriate means of regulating exposure.[14] Labeling can inform consumers (and workers who must use solvents) of the risks associated with using a substance and of actions that can be taken to reduce those risks, such as using the product in a ventilated area in the case of paint strippers. Consumers are free to decide if the benefits of using the product outweigh its risks combined with the costs of any risk-reducing actions.

---

[11]The U.S. EPA (1991) proposed a permit trading scheme for distributors of VOC-emitting products that presumably would pertain to a subset of the chlorinated solvents—TCE and PERC. Under their proposed scheme, distributors would be issued permits allowing them to sell a certain quantity of VOC-emitting products and then they would be allowed to trade permits among themselves. For partially reclaimable products such as solvents used in cleaning applications, the permits would reflect the difference between solvent sold and spent solvent accepted for recycling. This scheme would raise the demand for recycled solvent and thereby raise the prices, but it might have the effect of reducing the level of internal recycling below the optimal amount.

[12]The permit trading scheme proposed by the U.S. EPA (1991), although broader in scope than a trading scheme applied strictly to solvents, would be less costly to administer than trading among solvent users because the universe of potential traders is much smaller than the total number of users.

[13]The SCAQMD permit trading scheme covers more smog precursors than just PERC and TCE, since it is intended to reduce total emissions of all VOCs. Plans are in place to expand the scope of this VOC permit trading program to include a larger group of small VOC emitters. Regulators hope that by regulating a wider range of sources through trading, they will be able to avoid the costs of promulgating new command-and-control regulations. The flexibility of this program for solvent users may be limited, since regulations that cap solvent emissions as air toxics may not be violated under trading.

[14]Recall that methylene chloride is not a smog precursor and, according to the World Meteorological Organization (1985), it does not contribute to stratospheric ozone depletion. Given that METH is oxidized by exposure to hydroxy radicals in the atmosphere and breaks down into naturally occurring chemicals and that it also breaks down in groundwater, it may not pose a serious threat to third parties.

The welfare implications of product labeling are illustrated in figure 2-4. The demand curve for uninformed consumers is given by the curve D^U, which shows the marginal private benefits of using a solvent-containing product without acknowledging the potential risk associated with use. The demand curve under full information is given by D^I, which shows the marginal private benefits of consuming solvent-containing products taking into account the marginal risks to consumers associated with product use. The social welfare gains from informing consumers are given by the shaded triangle. Note also that consumer surplus abc is larger than if consumer solvents were banned or their availability was otherwise restricted to, say, Q^R.

In order for these welfare gains to be realized, product labels must be worded carefully. Labels should present an accurate assessment of the risks associated with product use and provide instructions as to how to use the product in such a way that risk is reduced. (For discussion of the challenges involved in labeling and understanding consumers' responses to labels, see Magat, Viscusi, and Huber [1988].) Labels should fully inform consumers about risks, but should not unnecessarily alarm them. (A more extensive discussion of the economics of labeling is included in chapter 3.)

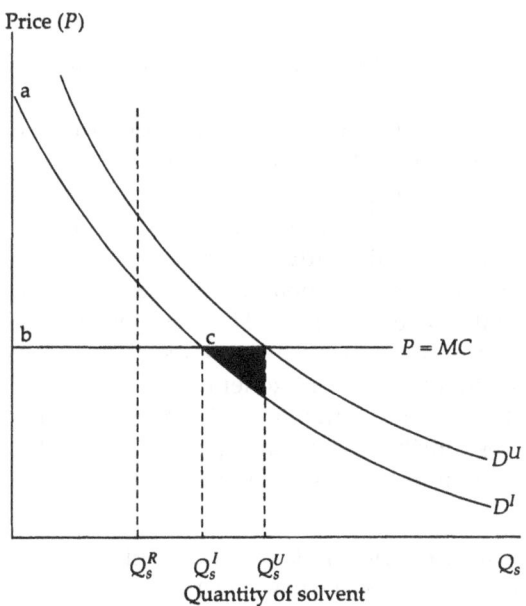

Figure 2-4. Welfare gains from product labeling.

An alternative to product labels would be the establishment of uniform product standards, such as measures limiting the solvent content of certain solvent-containing products or others that could reduce consumer exposure. Standards may be politically more attractive than labeling, particularly with consumers concerned about health effects, since standards take a harder line on reducing risks. Standards also may be preferable to labels when consumers are not well informed about their sensitivity to a product and, therefore, not necessarily able to take the steps necessary to achieve the appropriate level of exposure reduction. Manufacturers may also prefer standards, since they could reduce the need for public acknowledgment to potential customers that use of the product could be harmful. Indeed, manufacturers have been known to voluntarily institute risk-reducing product-content standards and may advertise these efforts as acts of public goodwill. Standards could potentially backfire if the consumer response is not consistent with the goal of the standard. For example, a standard that limited the METH content of a paint remover (or required an additive to reduce the volatility of the METH) might also limit the product's effectiveness and, therefore, lead consumers to purchase additional paint remover, leading to potentially greater harmful emissions than would have occurred with the stronger formulation of the product.[15]

Deposit-Refund System for Recyclers/Waste Handlers

Harmful exposure to chlorinated solvents can arise through improper disposal of spent solvent that results in groundwater contamination. Existing regulations under RCRA are designed to eliminate these exposures by banning the land disposal of spent solvent as well as the still bottoms remaining after recycling.[16]

RCRA regulations also mandate a manifest-based system for tracking solvent waste transport. The manifest system requires solvent users (and waste transporters) to keep written records of what happens to solvent waste from generation (acceptance for transport) through final disposal. Generators of waste containing chlorinated solvents are responsible for ensuring that waste is transported by a licensed individual and disposal occurs at a licensed facility. These

---

[15]The potential for this effect is limited by the cost of additional paint remover and the price elasticity of demand for paint remover.

[16]Under this regulation, solvent must be incinerated at a licensed hazardous waste incinerator or treated and disposed of at a RCRA hazardous waste landfill.

regulations have raised the cost of proper solvent disposal. At the same time, sustained high prices for virgin solvent have helped maintain a high level of demand for recycled solvent and, therefore, high demand for spent solvent for recycling. Almost all spent solvent that is not internally recycled is sold to solvent recyclers.[17] Thus, solvent users currently have no incentive to dump spent solvent, because it is a valuable commodity.[18] Solvent recyclers, on the other hand, may have an incentive to dump solvent still bottoms, given the high cost of incineration.[19]

To remove this incentive and avoid the social costs of illegal disposal, a deposit-refund system might be imposed on all parties accepting spent solvent for recycling and disposal of residuals. Under this system, the solvent waste handler would be required to pay a deposit to the government for every pound of spent solvent accepted for recycling (or still bottoms accepted for disposal). This deposit would be refundable in exchange for proof of recycling or proof of legal disposal at a licensed hazardous waste incinerator.[20]

The static effects of such a regulatory scheme on social welfare and on the solvent recycling industry are illustrated in figure 2-5. This diagram shows the market for recycled solvent and the marginal costs of solvent disposal for a given quantity of spent solvent $\overline{S}$.[21]

---

[17]Personal communication from Katy Wolf, Institute for Research and Technical Assistance, Los Angeles, July 22, 1991. Pekelney (1990) finds that nearly 80 percent of all category 211 wastes (liquid halogenated organic wastes, including METH, PERC, TCE, and TCA) tracked under the California Hazardous Waste Manifest System in 1988 were recycled. The quantity of waste identified as recycled in 1988 accounted for 93 percent of the waste for which a waste-handling method was identified.

[18]This incentive could change if solvent prices fall; however, given the long-term stability of real solvent prices and the high cost of solvent disposal, such a price drop is unlikely.

[19]Solvent users do have some incentive to seek recyclers who will not engage in illegal disposal, since such activity on the part of the recycler will require incomplete or forged Hazardous Waste Manifest forms and could result in eventual Superfund liability for the solvent user.

[20]The proposed deposit-refund system is analogous to an environmental bond, as discussed by Bohm and Russell (1985). Under the proposed system, the provider of waste-handling services essentially is required to post a bond sufficient to cover the cost of environmental cleanup that might be necessary as a result of its activities. When a batch of spent solvent is fully and appropriately processed, the bond is refunded. This form of deposit-refund is similar to the financial assurance requirements for hazardous waste treatment, disposal, and storage facilities under RCRA that require those facilities to carry insurance to cover damage from sudden accidental releases of substances.

[21]This solvent is left over from the previous period. The effects of the proposed

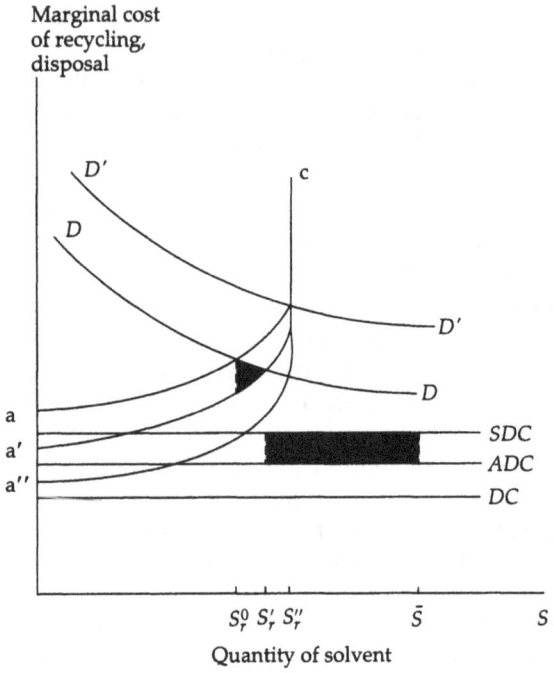

Figure 2-5. **Welfare and substitution effects of a deposit-refund system.**

The initial supply or marginal cost curve for recycled solvent is given by the curve ac. This curve turns vertical at the point S_r'', which is the maximum feasible amount of recycled solvent that can be extracted from a stock of spent solvent equal to $\bar{S}$. The marginal cost curve ac is the marginal cost of recycling minus the avoided marginal cost of solvent disposal. The marginal cost of cheap disposal (dumping) is given by the horizontal line labeled DC (disposal cost). According to the simple model underlying this diagram, disposal costs are accrued for all spent solvent in the possession of recyclers and then "returned" for that portion of the spent solvent that is recycled.[22] The constant marginal social cost of cheap (or "illegal") disposal is given by SDC.

---

deposit-refund system on generation of spent solvent in future periods is discussed heuristically in footnote 23, below.

[22]This assumption is consistent with the formulation of the cost of processing $\bar{S}$ as $RC(S_r) + DC \cdot (\bar{S} - S_r) = RC(S_r) - DC \cdot S_r + DC \cdot \bar{S}$ where RC is the recycling cost function, DC is the constant marginal cost of solvent disposal, and S_r is the quantity of solvent that is recycled. With this representation, the marginal processing cost of S_r is $RC'(S_r) - DC$.

In figure 2-5, market demand for spent solvent, given by DD, intersects the marginal cost curve, ac, at a point that leads to recycling of quantity S_r^0 from total spent solvent stocks of S_0. However, the socially efficient level of recycling, given the selected method of disposal, is S_r'', reflecting the intersection of the recycling cost curve adjusted for the avoided social cost of recycling, $a''c$, and the recycled-solvent demand curve.

Another method of solvent disposal with no externality cost, referred to as appropriate disposal, has a marginal cost given by ADC (appropriate disposal cost), which falls between SDC and DC. If all recyclers use this form of disposal for still bottoms, then the supply curve for recycled solvent becomes $a'c$ and a quantity equal to S_r' will be recycled. This level of solvent recycling is socially efficient (assuming no external costs associated with the activity of recycling itself) because the avoided social costs of the selected method of disposal are recognized in the decision about how much to recycle.[23] When appropriate disposal is used, waste handlers will dispose of quantity $\bar{S} - S_r'$ and no external costs will be incurred. The social welfare loss associated with using cheap disposal is given by the sum of the shaded triangle between ac and $a'c$ and the shaded rectangle between ADC and SDC.

If the demand curve is given by $D'D'$ instead of DD, then the equilibrium quantity to be recycled is S_r'' under all disposal schemes, and the social welfare loss of using cheap disposal is just $(\bar{S} - S_r'') \cdot (SDC - ADC)$.[24]

Imposing a deposit-refund system on solvent recyclers could encourage them to use the most socially efficient method of disposal.[25] The deposit-refund would be set equal to the level of mar-

---

[23]This analysis is static and does not consider the effect of internalizing disposal externalities on the future generation of spent solvent. As disposal costs rise, *ceteris paribus*, the price offered for spent solvent falls. This reduced price will have a positive effect on the level of internal recycling (because of lowered opportunity cost), although the higher cost of disposing of internally generated still bottoms may mitigate this effect somewhat. The higher disposal costs will cause an upward shift in the cost of virgin solvent and reduce the demand for solvent.

[24]Some industry experts believe that, given current disposal regulations and the high price of virgin solvent, demand for recycled solvent is currently in such a position and further disposal regulation will do nothing to increase recycling activity.

[25]Others (IEI, 1989; U.S. EPA, 1991) have proposed that a deposit be imposed on initial sales of virgin solvent with a refund to be issued when solvent is returned for recycling. The purpose of such a deposit-refund scheme would be to encourage solvent recycling and discourage dumping of spent solvent by solvent users. The deposit-refund system proposed here extends the incentive system to cover the disposal activities of recyclers and other solvent waste handlers.

ginal damages, or $(SDC - DC)$, times the total quantity of spent solvent accepted for disposal/recycling. The deposit would be refunded on all solvent that is either recycled or properly disposed of at a licensed hazardous waste disposal facility. Such a deposit-refund system would result in quantity S'_r of solvent being recycled and in proper disposal of the remaining solvent and still bottoms $(\overline{S} - S'_r)$. Assuming no solvent is lost as emissions during recycling, the firm will receive its entire deposit back $(SDC - DC) \cdot \overline{S}$ and will pay $ADC \cdot (\overline{S} - S'_r)$ for disposal.

The deposit-refund system raises the costs of illegal disposal and may drive some recycling/waste-handling firms out of business if the extra costs associated with legal disposal exceed the profits from recycling. This possibility is illustrated in figure 2-6. The recycler depicted in this graph is particularly inefficient at recycling, able to reclaim less than 60 percent of all spent solvents accepted for recycling, and must dispose of the residual. The curve labeled S gives the marginal cost curve of recycling (this time *not* net of avoided disposal cost), and DC and ADC are the cost of dumping and appropriate disposal, respectively. If the cost of disposal rises from DC to ADC, this recycler will be unable to cover its costs at the market

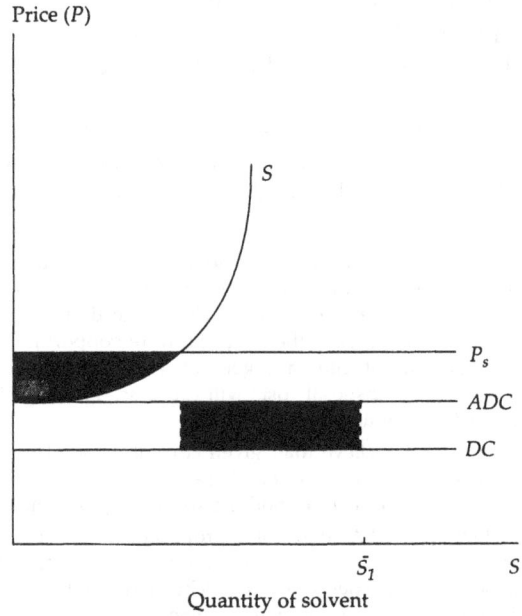

Figure 2-6. Deposit-refund system that drives inefficient recyclers out of business.

price for recycled solvent and will go out of business if area *A* exceeds area *B*. The proposed deposit-refund system raises the cost of illegal disposal and rewards appropriate disposal. This system also encourages recyclers to prevent air emissions since they will not receive a refund on any emitted solvents. The system is self-financing and, because, contrary to the examples used, some emissions will occur during recycling, the deposit-refund system will yield some net revenue to the government, which can be used to cover some of the costs of administering the deposit-refund program.

The administrative costs of this program could be high because of the number of solvent recyclers/waste handlers (less than 100) dispersed throughout the country.[26] The administrative burden of this program could be reduced by incorporating the initial deposit into the product tax proposed in the section above on Excise Tax on Sales of Solvents, as previously proposed by IEI (1989) and by the U.S. EPA (1991). This expanded product tax would incorporate damages from air emissions as well as from inappropriate disposal. Refunds would reward recycling and appropriate disposal by solvent users and spent solvent processors. Refunds would be set to equal the avoided damages of illegal disposal.

Given the good incentive properties of the deposit-refund system, there may be a reduced need for enforcement expenditures to ensure compliance under existing RCRA regulations. If the problem of illegal disposal of spent solvent and solvent still bottoms is serious, then the gains from instituting a deposit-refund program will justify the administrative costs. The severity of the illegal disposal problem should be further investigated prior to implementing the proposed deposit-refund system in order to determine the appropriate setting for the deposit (and whether or not the potential environmental damages avoided under such a program outweigh the costs).

Under the assumption that the benefits of solvent incineration outweigh the cost differences between incineration and other disposal methods and that the potential for illegal disposal under existing regulations is high, the deposit-refund mechanism is the only available incentive regulation that will encourage proper disposal. Disposal taxes will add to the incentive for illegal disposal. Increased enforcement of the existing land disposal ban is costly and may not be the most efficient way to increase compliance.

A possible ancillary benefit of the deposit-refund program out-

---

[26]The National Chemical Recyclers Organization reports that approximately 100 solvent recyclers are in business in the United States, but only a subset of these recycle chlorinated solvents.

lined above is increased recycling of spent solvent.[27] Assuming that some recycling increases are technically possible, another way to achieve these gains might be to institute a recycled-content requirement with tradable recycling credits. Under such a scheme, the regulator would set a minimum-recycled-content requirement that would have to be met for each batch of solvent sold. Under this regulation, solvent producers would be required to offer a certain quantity of recycled solvent for each pound of virgin solvent sold or to hold credits indicating that the requisite amount of recycled solvent had been produced by another bona fide recycler. Recycling credits for sale to virgin solvent producers and importers would be issued by the government for each pound of solvent reclaimed by recycling. These permits would be tradable, and a tightening of the recycled-content requirement over time would increase their value.

In addition to the likely technical infeasibility of obtaining substantial increases in recycling, the major problem with such a scheme is determining the appropriate recycling level. The credit market might ensure that the most efficient agents are performing the recycling (although this is not clear if, by increasing the demand for spent solvent, the credit market causes external recycling to be inefficiently substituted for in-house recycling), but it does not ensure that the efficient level of recycling will be performed. Given that the regulator may pick an inefficient or unobtainable level of recycling, a scheme that leaves that decision up to the market might be better advised. Also, if demand for solvents were growing (which it is not) or fluctuating significantly, the stock of spent solvent available for recycling might not be sufficient to meet a particular recycling target.[28]

CONCLUSIONS

The appropriate incentive mechanisms for regulating use and disposal of chlorinated solvents depend primarily on three considera-

---

[27]Katy Wolf of the Institute for Research and Technical Assistance believes that the possibility for increased recycling has been exhausted and that no increases in recycling activity will be forthcoming regardless of the amount of solvent regulation.

[28]Some of this uncertainty regarding the appropriate recycling target could be accommodated if the permit scheme were made more flexible by allowing a fee on virgin production in excess of allowed proportion and a subsidy for recycling in excess of the goal. This form of flexibility is proposed by Roberts and Spence (1976) as a means of improving the efficiency of a permit trading scheme in the presence of uncertainty about emission control costs.

tions: the extent of third-party effects, the extent to which dissipation of the solvent can be controlled during use, and the existing regulatory regime. The structure of the solvent market (number of producers, distribution channels, and number of consumers) is an important consideration in targeting regulations in a manner that maintains the appropriate incentive effects while minimizing administrative costs.

For solvent applications where third-party effects are small, such as consumer use of paint remover, the use of labels that warn about risks and instruct on risk-averting behavior during product use may be an efficient means of regulation. For solvents that dissipate completely during use and have significant third-party effects, a product tax imposed on the manufacturer for sales to dissipative applications will be an appropriate mechanism for reducing emissions as well as increasing efficiency in use.

If the solvent is partially containable, as in most cleaning applications, then eventually some form of disposal will be required for spent solvent or solvent still bottoms from recycling. For these situations, a deposit-refund may be desirable to ensure that proper disposal of residual solvent occurs.

Under current RCRA regulations regarding disposal of chlorinated solvents and associated conditions in the market for reclaimed solvent in the United States, solvent users face a high demand for spent solvent. Therefore, solvent users have a strong incentive to return all spent solvent to recyclers. However, the high cost of disposal of solvent still bottoms may result in some illegal disposal of still bottoms by solvent waste handlers and recyclers. Depending on the potential magnitude of the illegal disposal problem and associated environmental harms and the costs of implementing the regulation, a deposit-refund program imposed on solvent waste handlers/recyclers could provide the appropriate incentives to reduce illegal disposal. In the days prior to the RCRA land disposal ban of 1986, a deposit-refund scheme could have been instituted at an earlier stage in the solvent life cycle to encourage recycling by solvent users. By setting the appropriate level for a deposit-refund, such a program might have achieved the goal of the land disposal ban (appropriate disposal and increased recycling) at a lower cost.

Another important consideration in selecting incentive mechanisms is the toxicity of existing and potential substitutes for these substances. By considering the entire class of chlorinated solvents as a candidate for regulation, we include a large proportion of the potential substitutes to any particular solvent in a cleaning application. However, the life-cycle analysis in this chapter indicates that other substitutes are available in many applications and that some may be more harmful than solvents. In order to limit the possibility of regulation leading to increased harmful exposures, regulation must be broad enough in scope to deal with existing substitutes and flexible enough to anticipate the introduction and use of as-yet-unknown substitutes. An environmental bond, similar to the form of deposit-refund proposed above for solvent still bottoms, is one regulatory tool that could be used to discourage the introduction of harmful substitutes while limiting information requirements for regulators.[29]

---

[29]The use of such environmental bonds for regulating BFRs and potential substitutes are discussed in chapter 5 of this volume.

REFERENCES

ASTM (American Society for Testing and Materials). 1989. *Manual on Vapor Degreasing*, 3rd ed. (Philadelphia, PA: ASTM).

Bohm, P., and Russell, C. 1985. "Comparative Analysis of Alternative Policy Instruments." Pp. 395–460 in A. Kneese and J. Sweeney, eds., *Handbook of Natural Resource and Energy Economics* (Amsterdam: North Holland).

Chestnutt, T. W. 1988. *Market Response to the Government Regulation of Toxic Substances: The Case of Chlorinated Solvents*, N-2636-EPA (Santa Monica, CA: RAND Corp.).

Gaskins, P. 1974. "Alcoa Revisited: The Welfare Implications of a Second-Hand Market." *Journal of Economic Theory*, vol. 7, pp. 254–271.

HSIA (Halogenated Solvents Industry Alliance). 1989. *Chlorinated Solvents in the Environment* (Washington, DC: HSIA).

IEI (Industrial Economics, Inc.). 1989. *Multi-Media Incentives for Reducing Releases of SARA 313 Chemicals* (Cambridge, MA: IEI).

Magat, Wesley A., W. Kip Viscusi, and Joel Huber. 1988. "Consumer Processing of Hazard Warning Information." *Journal of Risk and Uncertainty*, vol. 1, pp. 201–232.

Mannsville Chemical Products Corporation. 1988. *Chemical Products Synopsis: Methylene Chloride* (Asbury Park, NJ: September).

———. 1989a. *Chemical Products Synopsis: Perchloroethylene* (Asbury Park, NJ: February).

———. 1989b. *Chemical Products Synopsis: Trichloroethylene* (Asbury Park, NJ: February).

———. 1990. *Chemical Products Synopsis: 1,1,1-Trichloroethane* (Asbury Park, NJ: October).

Martin, R. 1982. "Monopoly Power and the Recycling of Raw Materials." *Journal of Industrial Economics*, vol. 30, pp. 405–419.

Morton, P. R. No date. "Resource Recovery Through Solvent Reclamation." Mimeo.

Pekelney, D. 1990. "Hazardous Waste Generation, Transportation, Reclamation, and Disposal: California's Manifest System and the Case of Halogenated Solvents." *Journal of Hazardous Materials*, vol. 23, pp. 293–315.

———. 1991. *Analyzing Environmental Policies for Chlorinated Solvents with a Model of Markets and Regulations*. (Ph.D. dissertation, the RAND Graduate School).

Roberts, M., and M. Spence. 1976. "Effluent Charges and Licenses Under Uncertainty." *Journal of Public Economics*, vol. 5, pp. 193–208.

Tirole, J. 1988. *The Theory of Industrial Organization* (Cambridge, MA: MIT Press).

U.S. EPA (U.S. Environmental Protection Agency). 1991. *Economic Incentives: Options for Environmental Protection*. PM-220 (Washington, DC).

Wolf, K. 1990a. "Chlorinated Solvents: Sources Reduction in the Electronics Industry." Unpublished manuscript.

———. 1990b. "Chlorinated Solvents: Source Reduction in the Paint Stripping Industry." Unpublished manuscript.

———, and F. Camm. 1987. *Policies for Chlorinated Solvent Waste: An Exploratory Application of a Model of Chemical Life Cycles and Interactions*, R-3506-JMO/RC (Santa Monica, CA: RAND Corp.).

———, and C. Myers. 1987. *Hazardous Waste Management by Small Quantity Generators: Chlorinated Solvents in the Dry Cleaning Industry*, R-3505-JMO/RC (Santa Monica, CA: RAND Corp.).

World Meteorological Organization. 1985. *Atmospheric Ozone 1985—Assessment of Our Understanding of the Processes Controlling Its Present Distribution and Change*. Global Research Monitoring Project, Report No. 16, Vols. 1–3 (Geneva: WMO).

3

Formaldehyde

Like all of the substances discussed in the case studies of this book, formaldehyde is ubiquitous in the scope of its applications. However, a distinguishing feature of formaldehyde, apart from possible environmental and third-party exposure during production processes (by way of air and water pollution), is that few third-party effects appear to be associated with use of products containing formaldehyde. Rather, potential consumer exposure is limited to formaldehyde off-gases from formaldehyde-containing products (for example, furniture, cabinets, textiles, cleansers) in the home or workplace.

Two of our regulatory approaches are similar to approaches discussed in chapter 2, "Chlorinated Solvents": a combination of an excise tax on sales of formaldehyde to internalize third-party effects, and labeling to inform users about the potential for exposure in the home or workplace. In addition, we consider the setting of standards on off-gases from some products. We also consider the more complicated relationships governing market responses to labeling and standards when product liability insurance is available to producers or when product liability legislation applies to them.

The chapter begins with a review of the sources and uses of formaldehyde and a discussion of its adverse health and environmental effects. Next it traces the life cycle of formaldehyde from production through use, and then briefly reviews existing formaldehyde regulation.

As in the chapter on chlorinated solvents, these life-cycle and regulatory reviews set the stage for identifying regulatory incentives to reduce potential harmful exposures. We identify key features of formaldehyde production and use that have implications for selecting an appropriate regulatory mechanism, and then discuss candidate incentive-based approaches to regulating formaldehyde. To complement the discussion, we also review reasons why other approaches are less desirable. A concluding section summarizes the discussion.

SOURCES AND USES

Formaldehyde is a chemical compound of carbon, hydrogen, and oxygen. In addition to synthetic formaldehyde, formaldehyde occurs naturally as a colorless gas with a pungent odor. It is also a product of combustion in motor vehicles, power plants, manufacturing facilities, incinerators, and petroleum refineries, and of photooxidation of other hydrocarbons in the atmosphere.[1] Most synthetic formaldehyde embodied in consumer products is sold to producers in the form of water solutions; the other principal forms of manufactured formaldehyde are flake or powder. Manufactured formaldehyde is produced by the dehydrogenation of methanol (specifically, the oxidation of methanol using a silver or iron oxide–molybdenum oxide catalyst). Formaldehyde presently represents a sizable share—about 25 percent—of methanol demand.

The major use of formaldehyde is in the form of urea- and phenol-formaldehyde resins (see table 3-1). Urea-based resins are used primarily as adhesives in pressed-wood products such as particleboard, plywood, chipboard, medium-density fiberboard, and paneling. These resins are also used in textiles and paper (for example, they impart permanent-press characteristics to clothing and add strength to paper towels). Although phenol-based resins have slightly different properties (including less potential to be irritating to the eyes, throat, and nose, as discussed below), they can be substituted for urea-based resins in some applications. However, supplies of phenol-based resins are limited, and at present they are significantly more expensive than urea-based resins. In addition, phenol-based resins are darker and so are aesthetically less desirable in certain applications (for example, phenols are generally used in outdoor rather than indoor wood products). Both categories of formaldehyde resins, in fact, consist of many additional types of resins, as a resin is typically tailored to the characteristics of the particular wood, pulp, or other inputs with which it is being used. Table 3-2 contains additional information about uses of formaldehyde resins

---

[1]This case study does not consider the regulation of formaldehyde that is emitted from combustion processes (so-called secondary formaldehyde). Krupnick, Walls, and Toman (1990) summarize some evidence that reduction of formaldehyde emissions from vehicle gasoline combustion would have little effect on air quality, although they note that formaldehyde emissions from an alternative fuel, methanol, are much greater. See also Machiele (1990) and Fishbein and Henry (1990). Secondary formaldehyde emissions from manufacturing and other industrial processes are now regulated as a hazardous air pollutant under the Clean Air Act Amendments of 1990 (see table 3-4 in this chapter).

Table 3-1. End Uses of Formaldehyde, 1990

Formaldehyde derivative	End uses	Percentage of formaldehyde output
Urea-formaldehyde resins and concentrates	Particleboard, plywood, other (textiles, paper, pesticides)	30
Phenol-formaldehyde resins	Plywood and particleboard, fiberglass, other (coatings, adhesives)	24
1,4-Butanediol	Engineering plastics	12
Acetal resins	Engineering plastics	9
Hexamethylenetetramine	Curing, explosives, bactericides	6
Pentaerythritol	Unknown	5
Melamine-formaldehyde resins	Unknown	4
Methyl diphenyl diisocyanates	Unknown	4
Miscellaneous	Unknown	6

Source: Mannsville Chemical Products Corporation, *Chemical Products Synopsis: Formaldehyde* (Asbury Park, NJ, July 1990).

in the wood-products industry, which accounts for about 55 percent of total formaldehyde demand.

In the United States, formaldehyde is produced by about a dozen firms at 46 plant locations. Because transportation costs are large compared with plant capital costs, plants are generally located near the producers that use formaldehyde derivatives in their products. For example, because of the significant demand for urea-formaldehyde resins for wood products, formaldehyde plants tend to be located near wood-product producers in the Pacific Northwest and the Southeast. Formaldehyde imports are negligible (although imports of formaldehyde-containing products can be larger, depending on the product); U.S. suppliers of formaldehyde tend to operate below capacity. The price of formaldehyde generally moves with that of methanol, and the demand for formaldehyde is reported to be driven by trends in the construction industry.

HEALTH AND ENVIRONMENTAL EFFECTS

Some human exposure to formaldehyde occurs from natural concentrations in the body, and from small, naturally occurring background

Table 3-2. Industry Status for Urea- and Phenol-Formaldehyde Resins

Feedstock	Production	Wood products and primary uses
Formaldehyde made from methanol by silver-catalyzed or metal-oxide process	Urea-formaldehyde resin (30% of formaldehyde output)	*Particleboard*, used in furniture, fixtures, cabinets, and floor underlayment
		Medium-density fiberboard, used in furniture, fixtures, cabinets, and floor underlayment
		Hardwood plywood, used in furniture, cabinet doors, and wall paneling
	Phenol-formaldehyde resin (25% of formaldehyde output)	*Hardboard*, used in wall paneling, exterior siding, appliances, some furniture
		Waferboard, used in hardboard, crafts, pallets, fixtures
		Softwood plywood, used in wall and roof sheathing, concrete forms, subfloors, some furniture
		Particleboard, used in millwork, boat building, packaging

Note: Chemical substitutes for urea-formaldehyde (UF) resins include phenol-formaldehyde (PF) resins, isocyanates (ISOs), and polyvinyl acetates (PVAs). PF resins are more costly than UF resins and are dark-colored and thus are not suitable for interior use. ISO is very costly and tends to bind with the metal of the press faces, which causes wear on the equipment. PVAs cannot be used for particleboard because they deform under stress. Off-gases from PF resins can be less than those from UF resins by factors of 10 to 20, but products made with PF resins require longer pressing times and higher temperatures for processing, resulting in higher workplace emissions of volatile organic compounds from wood. ISO can cause irritation and immunologic sensitization of the respiratory tract.

Sources: Mannsville Chemical Products Corporation, *Chemical Products Synopsis: Formaldehyde* (Asbury Park, NJ: September 1988). Personal communication from William H. McCredie, Executive Vice President, National Particleboard Association. Presentation notes in the form of a letter, dated May 9, 1991, to Mark Greenwood, Director, U.S. Environmental Protection Agency's Office of Toxic Substances, from John F. Murray, President, Formaldehyde Institute; William H. McCredie, Executive Vice President, National Particleboard Association; and William J. Groah, Technical Director, Hardwood Plywood Manufacturers Association. John Bower, *The Healthy House: How to Buy One, How to Cure a "Sick" One, How to Build One* (New York, NY: Carol Communications, 1989), pp. 151, 153.

amounts in the air and in fruits, vegetables, and dairy products. Further exposure also can come from the motor vehicle and industrial combustion processes noted earlier. In the home, sources include vapors from wood products (for example, furniture, floor underlayment, cabinets, paneling) containing formaldehyde resins and from cigarette smoke, gas stoves, heaters, and the breakdown of cooking oil as it is heated. Additional sources include shampoo, toothpaste, vaccines, fertilizer additives, and household disinfectants.

As in the case of some of the chlorinated solvents, indoor air exposures are limited to workers or consumers affected by formaldehyde emissions from formaldehyde-containing products. In addition, harmful exposures to third parties can occur, principally further upstream in the production of formaldehyde, by way of producer discharges. Examples are water pollution from the disposal of distillation bottoms during the production of formaldehyde and waste water during the production of formaldehyde-based insecticides.

Formaldehyde vapors can irritate the eyes and skin, and inhalation of the vapors can affect the respiratory system. Relatively low concentrations have caused asthmatic symptoms in some individuals. Formaldehyde is known to cause nasal tumors in some rodents, but according to the American Medical Association (Council on Scientific Affairs, 1989), evidence of formaldehyde carcinogenicity in humans is inconclusive. However, on the basis of health risk assessments, the U.S. Environmental Protection Agency (EPA) now classifies formaldehyde as a probable human carcinogen, and the Occupational Safety and Health Administration (OSHA) has set standards for workplace exposure limits and presently is reviewing new standards.

Vapors from high concentrations of formaldehyde have a noticeably disagreeable odor that can signal the need for an individual to ventilate his or her environment or take other measures to reduce exposure. At lower concentrations, typical of most household products containing formaldehyde, the odor is less noticeable (but may still be irritating). The odor is most noticeable when the product is new, as the off-gases, especially from pressed-wood products containing formaldehyde-based resins, decrease over time. The rate of decrease varies with the type of resin (for example, vapors from phenol resins decline more rapidly than those from urea resins); estimates in the literature range from a few months to a few years.[2]

LIFE CYCLE IN VARIOUS APPLICATIONS

Table 3-3 outlines the stages of formaldehyde production and use and the nature of exposure during those stages. Third-party exposure

---

[2]For example, see test results included in briefing materials submitted at the National Particleboard Association/Consumer Product Safety Commission Meeting, January 19, 1990 (mimeo, available from authors); Friess (1991); and Groah and Gramp (1991).

Table 3-3. Formaldehyde Life Cycle and Sources of Exposure

Stage of production or use	Exposure
Feedstock	Worker, ambient air, water
Distribution of feedstock	Worker, ambient air, water
Production of formaldehyde-containing product	Worker, ambient air, water
Consumption (use) of product	Consumer (user)
Servicing or repair of formaldehyde-containing products	Little or none
Disposal	Little or none (biodegradable)

occurs principally during upstream production and includes occupational exposure and environmental exposure to air and water. Further downstream, during use of formaldehyde-containing products, the health effects noted in the preceding section can arise from off-gases. The use of recycled building materials containing formaldehyde (for example, pressed-wood products such as cabinets and flooring) that have been converted to wood chips or mulch for gardening or animal bedding may be restricted. However, such restriction is less related to environmental or health reasons than it is to the difficulty of chipping or mulching products containing resins.

EXISTING STATUTES AND REGULATIONS

Numerous statutes and regulations govern various aspects of formaldehyde production and use. Table 3-4 lists the most significant of these, and also notes examples of state regulatory efforts (California's Proposition 65) and voluntary industry standards for off-gases from some wood products. Two questions are suggested by this legislation: (1) What are the determinants of industry's willingness to set standards voluntarily? and (2) How have consumers responded to formaldehyde labeling under Proposition 65 and under the U.S. Department of Housing and Urban Development (HUD) requirements? Both topics are discussed further in the section on Proposed Incentive Mechanisms, below.

FORMALDEHYDE ATTRIBUTES RELEVANT TO ECONOMIC INCENTIVE DESIGN

As emphasized throughout the case studies in this book, we assume that the goal of toxic substance regulation is to reduce potentially

Table 3-4. Statutes and Regulations Pertaining to Formaldehyde

Type of action/ responsible agency	Statute or regulation	Uses or sources of exposure covered
FEDERAL		
Consumer Product Safety Commission		Does not mandate standards, but asks for voluntary consensus process.
Occupational Safety and Health Administration	Occupational Safety and Health Act	Sets exposure limits of 1 part per million (ppm) in the workplace (weighted average).[a]
U.S. Department of Housing and Urban Development (HUD)		Requires formaldehyde emission standard for products used in manufactured-home construction. Requires irritation notice (not carcinogen labeling) in manufactured homes (kitchen label and insertion in home guide). For urea-formaldehyde resin, HUD standard differs by product type (hardwood plywood, particleboard) for manufactured housing.
U.S. Department of the Treasury	Excise Tax of 1991	Taxes formaldehyde imports.
U.S. Environmental Protection Agency (EPA)	Clean Air Act Amendments	Regulate formaldehyde as a hazardous air pollutant under Title III. Set standards under section 243 for emissions from light-duty clean-fuel vehicles (essentially, formaldehyde emissions from methanol vehicles).
EPA	Resource Conservation and Recovery Act	Regulates wastes of formaldehyde in generation (i.e., during the production of formaldehyde products), transport, treatment, and disposal.
EPA	Superfund Amendments and Reauthorization Act	Under Title III (Emergency Planning and Community Right to Know), sets reporting requirements for released amounts.
STATE		
	California Proposition 65	Requires businesses in California to warn consumers and workers of dangers of exposure.

Table 3-4 (*continued*)

Type of action/ responsible agency	Statute or regulation	Uses or sources of exposure covered
VOLUNTARY INDUSTRY ACTION		Sets voluntary standards for formaldehyde emissions from the wood-products industry as follows[b]: 0.3 ppm for medium-density fiberboard, 0.3 ppm for particleboard. Warning labels often stamped on products.

[a]The American Council of Governmental Industrial Hygienists has proposed that a more stringent level of 0.3 ppm be established.

[b]Parts per million under specific test conditions for exposed surface area per unit time, given air exchange and relative humidity.

Sources: Mannsville Chemical Products Corporation, *Chemical Products Synopsis: Formaldehyde* (Asbury Park, NJ: September 1988). Internal Revenue Service, *Excise Taxes for 1991* (Washington, DC: U.S. Department of the Treasury, 1991). National Particleboard Association, *American National Standard for Wood Particleboard, and Medium Density Fiberboard* (Gaithersburg, MD: February 1989). John Bower, *The Healthy House: How to Buy One, How to Cure a "Sick" One, How to Build One* (New York, NY: Carol Communications, 1989). National Particleboard Association, Memorandums to Members, Jan. 2, 1991, April 12, 1991.

harmful exposures. The life-cycle analysis of formaldehyde indicates that potentially harmful exposures can occur during manufacture and end use. Exposure during manufacture can affect society at large, whereas exposure during the use of formaldehyde-containing products is confined primarily to the immediate user(s) (for example, consumers of formaldehyde-containing products, or office workers whose furniture may contain formaldehyde resins).[3] In this regard, formaldehyde is different from cadmium, which has pronounced third-party effects (see chapter 4), but similar to some applications of chlorinated solvents (particularly methylene chloride in household paint strippers), which produce few third-party effects at this stage of the life cycle.

As in the cases of chlorinated solvents and cadmium, the potential for exposure downstream from feedstock production (where there are just a few producers) proliferates rapidly, from hundreds of producers of formaldehyde-containing products to millions of end users of these products. Accordingly, intervention upstream in the

---

[3]Again, we reiterate that our focus in this case study is on emissions from formaldehyde products, not formaldehyde emissions from combustion processes.

life cycle is likely to be administratively easier. However, the effect of upstream intervention on downstream uses is difficult to predict. For example, the bulk of a general tax on formaldehyde purchases by resin producers would be passed on to the customers least able to substitute other products. Although this has the intended result—reduction of formaldehyde demand—the customers whose demand falls the most may not be those whose products are most harmful. Because formaldehyde resins are generally tailored for specific applications (for example, for use with various types of wood and textiles), it may be possible to vary a general tax rate by type of resin, but again, administering and enforcing a highly differentiated tax may not be cost-effective. It might be administratively feasible to set somewhat higher taxes on urea-formaldehyde resin than on phenol-formaldehyde resin, however, given their different off-gassing characteristics. This could encourage an increased supply of phenol (presently very limited) or the addition of off-gas barriers (for example, varnishes, laminates) to end products.

The absence of third-party effects at the end-product stage suggests that the most appropriate form of regulatory intervention would be to provide more information to users about health effects and to identify possible mitigating actions that users might take. Provision of this information could take several forms, including product labels to apprise users of effects of exposure and to list actions to reduce exposure.

Providing more information might be particularly effective in the case of formaldehyde products for several reasons:

• There is evidence that different types of users are more sensitive than others. Thus, information could protect sensitive users without restricting the choices of less-sensitive users.

• There are mitigating actions that users can take if they are informed, such as ventilating the home or workplace or controlling humidity.

• Off-gassing varies with the age of the product; thus, indoor air emissions decline over time.

The setting of industry standards for off-gases from end products is another form of information that might be useful. However, as the next section suggests, the best form for information (for example, how a label should be worded or the level at which a standard should be set) is far from clear, but it is critical to the effectiveness of information as a regulatory mechanism.

PROPOSED INCENTIVE MECHANISMS

As suggested by the case studies in this book, the life-cycle approach to understanding the toxic substances that we consider implies the desirability of multiple interventions. In the case of formaldehyde, two levels of intervention are suggested: one level to internalize third-party effects upstream and one to address user exposure downstream. Table 3-5 summarizes various regulatory options and their advantages and disadvantages at different life-cycle stages. The sections below elaborate on this summary table.

Internalizing Third-Party Effects: Product Tax or Tradable Permit

An appealing incentive-based approach to limit formaldehyde emissions into air and water during production processes is a product tax that equals the marginal environmental and health damages associated with the emissions. Because the number of primary producers of formaldehyde is fairly small compared with the number of intermediate producers who use formaldehyde in their products, a tax levied on primary producers may be administratively easier than a tax levied on intermediate producers. The disadvantage of a tax levied on primary producers, however, is that it will be passed forward more heavily to applications for which there are few substitutes rather than to applications whose environmental and health effects are largest.

On the other hand, if these higher-effect applications can be identified, and if evading the tax is difficult (for example, low-taxed formaldehyde derivatives are not available for substitution), then taxes could be varied according to application to reflect differences in damages. In this case, the taxes would be levied at the point of sale to the intermediate producer rather than on the primary manufacturer. Because most formaldehyde is used as a solution in water, it is costly to ship long distances. Thus, it might be feasible to tailor point-of-sale taxes to geographically specific damages. If damages are difficult to calculate, then a standards-and-taxes approach could be taken, with emission standards set to reflect local environmental conditions.

The effects of a tax follow the diagram in figure 2-1. Figure 3-1 extends that diagram to show a spatially differentiated tax and the effects of transportation costs. Two production facilities are located at the origins of the axes in the figure (*PF*-1 and *PF*-2); *PF*-1 is assumed

Table 3-5. Regulatory Options for Formaldehyde

Exposure point	Regulatory option	Authors' verdict	Comments
PRODUCT MANUFACTURE			
Workplace	Labeling	Yes	No externality associated with use, and goal is better-informed user.
			Permits exposure mitigation; may include standards.
	Emissions or product tax	Maybe	May reduce use but labeling is also needed to protect workers, including those most sensitive.
	Tradable permits	Maybe	May reduce use but labeling is also needed to protect workers, including those most sensitive.
	Deposit-refund	N.A.	
	Command and control	Maybe	Might be advised, especially in the form of standards, if information markets fail under labeling.
Third parties	Labeling	N.A.	
	Emissions or product tax	Yes	May be administratively easier if applied to primary producers, although tax may be forwarded to applications with fewer substitutes rather than greater risk.

Tradable permits	Yes	May be administratively easier if applied to primary producers, although may be used in applications with fewer substitutes rather than greater risk.
Deposit-refund	N.A.	
Command and control	Maybe	Might be advised if tax or permits are too costly to administer.
PRODUCT USE		
Labeling	Yes	No externality associated with use, and goal is better-informed user. Permits exposure mitigation; may include standards.
Emissions or product tax	No	May reduce use but labeling is also needed to protect workers, including those most sensitive.
Tradable permits	No	May reduce use but labeling is also needed to protect workers, including those most sensitive.
Deposit-refund	N.A.	
Command and control	Maybe	Might be advised, especially in the form of standards, if information markets fail under labeling.

Note: N.A. = not applicable.

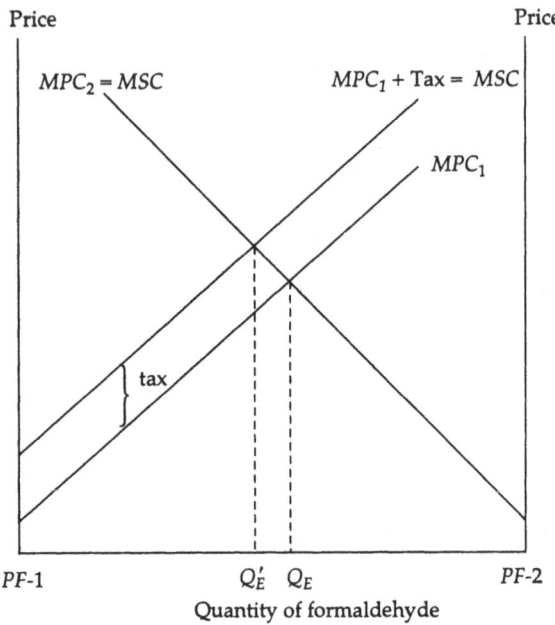

Figure 3-1. Spatial variation of a formaldehyde production tax.

to be located in an air-quality nonattainment area, and PF-2 is assumed to be located in an attainment area. The supply curves (MPC_1, MPC_2) slope upward with distance to show that transportation costs increase with distance from the production facility (MPC net of transportation costs is assumed constant). MPC_2 equals marginal social cost (MSC) in the attainment area. Quantity to the left of equilibrium output (Q_E) is supplied by PF-1 and to the right by PF-2. Thus, market boundaries are governed by transportation costs. If a tax reflecting MSC is levied on formaldehyde produced at PF-1, then, because of that region's nonattainment status, MPC_1 shifts upward by the amount of the tax. The effect is to reduce output from PF-1 and increase output from PF-2, as shown by Q_E'.

Alternatives to a product tax include emission taxes and tradable permits, which, like product taxes, ideally would be set to reflect damages associated with emissions. However, taxes and permit levels based on emissions alone present formidable monitoring challenges. A product tax generally offers incentives for emission reduction *and* has relatively lower administrative, enforcement, and compliance costs. To the extent that a product or emission tax or a tradable permit also increases the direct cost of formaldehyde production, the incentive may also confer ancillary benefits of reducing workplace exposure (see section in chapter 2 on Proposed Incentive Mechanisms).

A permit trading scheme that allowed producers to bid for the right to manufacture formaldehyde could help ensure that air- and water-quality standards were met, although the direct costs of permits are less predictable than tax rates (see, e.g., Weitzman [1974], Roberts and Spence [1976]). The quantity of permits made available, like tax rates, might also be regionally tailored to be more restrictive in air- or water-quality nonattainment regions. Permits could be issued and tradable within geographic regions surrounding the facilities producing formaldehyde, with the number of permits available set according to air- and water-quality standards relevant to the area. High costs of transporting formaldehyde would serve as a check on the movement of formaldehyde from low-permit-fee locations to high-permit-fee locales, provided the cost of a permit was less than the transportation costs (see figure 3-1, where the vertical distance between MPC_1 and MPC_1 + tax would represent the cost of a permit).

Labeling to Inform Consumers

Another concern associated with some formaldehyde products is exposure to consumers of the products in the home or workplace. Examples are formaldehyde off-gases from some wood products or textiles. In these applications, the potential third-party effects may be small, and therefore labeling (of domestic as well as imported products) may be an appropriate regulatory mechanism. It can inform consumers of the risks associated with using a product and of actions that can be taken to reduce those risks, such as using the product in a ventilated area or under climate-controlled conditions. Other alternatives including choosing formaldehyde-containing products that have barriers such as varnishes or veneers that can limit off-gasing, or substituting products that contain no formaldehyde. A labeling scheme would allow consumers to decide if the benefits of product use outweigh its risks (or the benefits of residual risk, after the consumer takes risk-reducing actions, combined with the costs of these actions).

The welfare implications of formaldehyde product labeling are illustrated in figure 3-2. This figure shows the effects of labeling when consumers vary in their sensitivity to a toxic substance or their

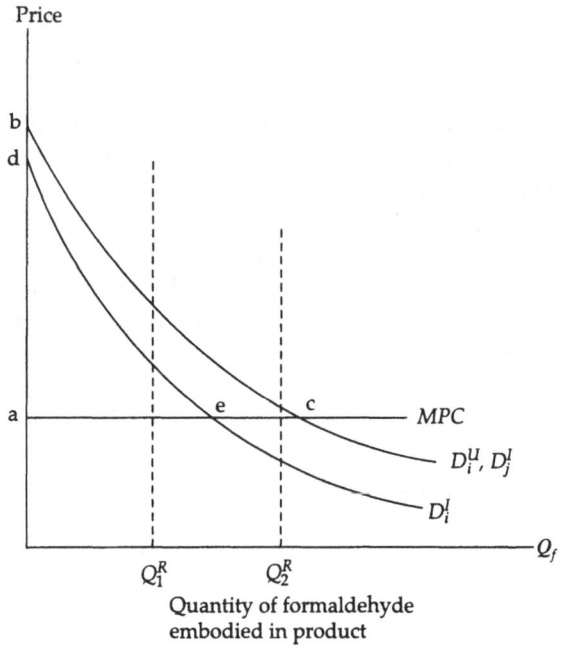

Figure 3-2. Possible effects of labeling versus output restrictions on consumer surplus.

opportunities to take risk-reducing actions.[4] The demand curves for two consumers, i and j, reflect uninformed (U) and informed (I) states of the world. Consumer i's behavior is affected by the label; consumer j's is unaffected. After labeling, consumer surplus is represented by the sum of triangles abc and ade. Thus, consumer surplus can be larger under labeling than under regulatory interventions (such as restrictions on formaldehyde content of or off-gases from products, or other mandated standards) that restricted output to Q_1^R to protect the most sensitive consumer or, more generally, that constrained the supply to Q_2^R.

The benefits of labeling compared with those of supply restrictions depend, however, on individual variation in sensitivity to off-gases. Although few studies on individual variation have been undertaken, the existing data imply that numerous factors (for example, indoor climate and ventilation, other sources of indoor pollution,

---

[4]This extended analysis applies as well to the discussion of labeling solvents in chapter 2.

and personal characteristics such as age, asthma, and smoking) could contribute to marked variation across the population. Furthermore, anecdotal evidence on the costs of product substitution suggests that they can be quite high (for example, the cost difference between solid and pressed-wood furniture). Thus, the size of the consumer surplus triangles are likely to be large for the general population, but may differ markedly for individuals. If so, labeling may be preferred to product bans or other restrictions (see section on Product Standards, below).

In some instances industry has voluntarily undertaken labeling of formaldehyde products.[5] Producers of particleboard and medium-density fiberboard have voluntarily set off-gas standards for their products. In the case of new manufactured homes (that is, prefabricated homes and house trailers) floor underlayments and cabinets in combination may emit more off-gases per unit of interior area than are emitted in other indoor settings. For this reason, the U.S. Department of Housing and Urban Development requires labeling to indicate that building materials emit formaldehyde (see figure 3-3). The label must be displayed in the kitchen of manufactured homes prior to sale as well as inserted in the consumer manual for the home.

Whether information is, in general, undersupplied for these or other formaldehyde products, and if so, whether government should encourage or require additional labeling depends on a host of factors. Little is available in the literature to shed light on the issue (see Ippolito [1984] for a discussion of this gap). However, figure 3-4 suggests one rationale for industry to label voluntarily (or to reformulate a product and label it reformulated). In panel A, the downward shift of the informed-consumer demand curve (D^I) results in output Q_2 and producer surplus of area abc. If the producer surplus could be increased by product labeling or reformulation, then these steps might be voluntarily undertaken. Panel B suggests that although labeling or reformulation might increase supply costs (MC shifts up to MC'), consumers respond by increasing demand somewhat. (They are willing to purchase more of the product than represented by D^I in panel A, because use can be safer, but $D^{I'}$ does not fully shift back to D^U because precautions must still be taken, at some cost [for example, ventilating], or because the reformulated product may be less effective.) As constructed in panel B, the small

---

[5]In addition, the Consumer Product Safety Commission has been encouraging industry to undertake additional voluntary labeling of some formaldehyde products.

IMPORTANT HEALTH NOTICE

Some of the building materials used in this home emit formaldehyde. Eye, nose, and throat irritation, headache, nausea, and a variety of asthma-like symptoms, including shortness of breath, have been reported as a result of formaldehyde exposure. Elderly persons and young children, as well as anyone with a history of asthma, allergies, or lung problems, may be at greater risk. Research is continuing on the possible long-term effects of exposure to formaldehyde.

Reduced ventilation resulting from energy efficiency standards may allow formaldehyde and other contaminants to accumulate in the indoor air. Additional ventilation to dilute the indoor air may be obtained from a passive or mechanical ventilation system offered by the manufacturer. Consult your dealer for information about the ventilation options offered with this home.

High indoor temperatures and humidity raise formaldehyde levels. When a home is to be located in areas subject to extreme summer temperatures, an air-conditioning system can be used to control indoor temperature levels. Check the comfort cooling certificate to determine if this home has been equipped or designed for the installation of an air-conditioning system.

If you have any questions regarding the health effects of formaldehyde, consult your doctor or local health department.

Figure 3-3. Formaldehyde irritation label required by the U.S. Department of Housing and Urban Development for manufactured homes.

shift in MC' plus the relatively larger shift in $D^{I'}$ increase the producer surplus (area *def* exceeds *abc*).[6]

This figure suggests only one possible result: clearly, much more analysis is needed to determine industry's willingness to label or otherwise respond to exposure issues. Other factors include, but are not limited to, how the information would be communicated (for

---

[6]The supply curve in figure 3-4 depicts the industry as imperfectly competitive. If the industry is more perfectly competitive, then producers may be less willing to label voluntarily because producer surplus areas would not be available for capture except in the short run or upon the offering of a different product, such as a reformulated one. For a related discussion, see the discussion of Tirole (1988, especially chapter 7) of the effects of different types of industry organization on product differentiation. See also the case study by Calfee (1986) of cigarette manufacturers' voluntary use of health warning labels in the early 1950s as a means of interbrand competition.

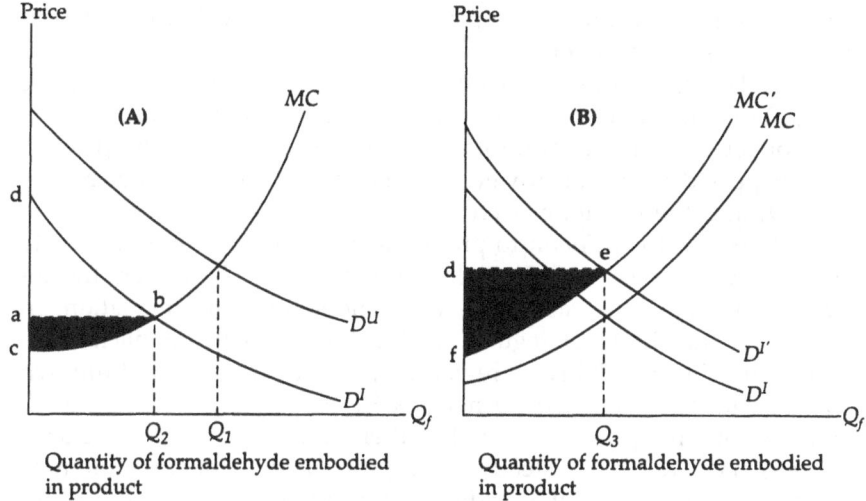

Figure 3-4. Producer incentives to label or reformulate formaldehyde-containing products.

example, the terminology as well as the tone of the wording on the label) and how it would affect consumer and producer behavior. Although no follow-up studies of the behavioral effects of formaldehyde labeling have been conducted, studies of other products (for example, cigarettes, insecticides, and cleansers) indicate that it is extremely difficult to predict consumer responses (see, e.g., Ippolito [1981], Ippolito and Ippolito [1984]; Magat, Viscusi, and Huber [1988]). This is the case whether the product merely produces irritation, such as of the respiratory tract, or whether it is or may be carcinogenic (both situations are relevant to formaldehyde). Ippolito (1981) notes that, typically, the demand response consists of a change in *quality* as well as a change in *quantity*; thus, substitution among formaldehyde products (given variation in off-gasing) as well as substitution away from these products might be expected.

An industry concern about warnings or other information on labels is that such information may unintentionally trigger liability claims that are meritless, that is, for health effects that are unrelated to a firm's activities or products. These claims may represent retrospective liability for past use of the product or be associated with current use. There is reason to believe that the mere communication of information can trigger socially inefficient litigation (see, e.g., Huber [1988], Huber and Litan [1991], and references cited in both volumes). This observation probably is related in part to the way in which communication takes place, given some evidence that consumers frequently overestimate highly publicized risks (see, e.g., Viscusi [1990], and references therein). If producer liability costs are large but the claims are meritless, then the social costs of litigation can offset the benefits of information. Figure 3-5 shows the producers' expected marginal private cost (*EMPC*) given the specter of liability, and the resulting decrease in output.

A priori, it is difficult to predict whether a triggering of liability would be the case with formaldehyde products. However, the industry points to the case of urea-formaldehyde foam insulation. In that case the threat of regulation, together with the publicity surrounding the possibility of regulation, led to the virtual shutdown of the industry (see also Posner [1986], chapter 3, for examples in the case of other products). On the other hand, sales of manufactured homes have remained steady since promulgation of formaldehyde labeling. Other anecdotal evidence indicates that California's Proposition 65 has led to some lawsuits (not related to formaldehyde), but not as many as many experts predicted (see, e.g., "Warning: Litigation Can Be Hazardous," 1990, p. 2403). A better understanding of these experiences and improved understanding of how the public processes information and perceives risk are overarching issues in

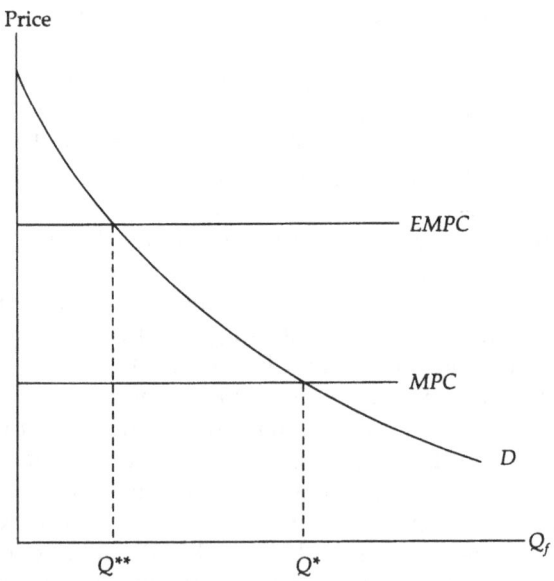

Figure 3-5. Possible effect of the specter of product liability on output.

pursuing labeling of formaldehyde products. As noted earlier, follow-up studies of consumer responses to formaldehyde labeling in manufactured homes and to changes in demand for wood products that are voluntarily labeled would be helpful in shedding light on this topic.[7]

Product Standards

Product standards can also serve to inform consumers of potential exposure risk. Standards can range from legislated minimum levels of product quality to guidelines that merely inform consumers about quality levels (for example, tar and nicotine content of cigarettes). In the case of formaldehyde, standards have been legislated for workplace emissions of formaldehyde during its production and in the manufacture of formaldehyde-containing products and for indoor emissions for manufactured homes (see table 3-4). As noted earlier, standards also have been voluntarily set by the wood-products industry for off-gases from medium-density fiberboard and particleboard.[8]

Standards have the benefit of providing consumers with a benchmark for behavior. Labeling without standards may not offer consumers enough guidance to choose among different products. An example is wood products: (a) some are close substitutes for each other but may have different emission characteristics or may include emission barriers such as varnishes or veneers; and (b) consumers are likely to vary in their sensitivity to emissions and in their willingness to take defensive actions such as climate control. In the case of many products, consumers' repeat purchases signal satisfaction with the quality of the products, that is, produce a reputation effect (see discussion in Phipps, Allen, and Caswell [1989]; see also Ippolito [1990], Costanza and Perrings [1990]). However, some formaldehyde-

---

[7]In the case of California's Proposition 65, which requires labeling on all products and services that present a significant risk of cancer, there is some evidence that a proliferation of labels has led consumers to shrug off almost all labels, since it appears that "everything is bad." Accordingly, it seems that providing relative-risk information may be useful. Consumers might be apprised of how a product's riskiness compares with other activities and with substitutes for the product, so that consumers do not substitute even riskier products for those with small risks. See discussion in Phipps, Allen, and Caswell (1989). These issues are at the forefront of risk communication research (for additional references and discussion, see Magat, Viscusi, and Huber [1988]; Viscusi and Hersch [1990]).

[8]These standards specify parts per million of formaldehyde per exposed surface area per unit time, given assumptions about typical air exchange and relative humidity.

containing products (for example, furniture, cabinets) are bought infrequently, and thus repeat buying is less of a check on the market. In the case of other more frequently purchased products, such as textiles or cleaning products, reputation effects may greatly influence the formaldehyde content of products if consumers associate symptoms of formaldehyde irritation with the product.[9] Thus, reputation effects may function to modify a firm's behavior, depending on the frequency of purchase of a product and the information available concerning it. Standards may be a desirable part of the information.

Standards can be difficult to set, however, especially given variations in indoor air pollution, climate, and ventilation, and the multiple sources of formaldehyde emissions from a given product (for example, both the wood and textile components of an upholstered product could be sources).[10] Moreover, if inappropriately set, standards can result in too much or too little product or product variation supplied to the market.

Given these concerns, broad standardized guidelines for formaldehyde emissions (along the lines of energy efficiency standards for appliances) rather than mandated minimum off-gas levels might permit a wider range of products to be supplied. Figure 3-2, described earlier, illustrates the problem of setting standards. The regulated levels Q_1^R and Q_2^R both miss the intersection of demand and marginal private cost (MPC). The asymmetry of information between regulator and consumer—the consumer, if informed of the potential for exposure, is more aware of his or her sensitivity to off-gases and willingness to take mitigating action—provides a strong argument for broad guidelines rather than highly specific standards. As demonstrated in figure 3-2, the disadvantage of specific standards is that they might significantly reduce consumer surplus. This will be the case particularly if the off-gas characteristics of products vary widely and if consumers vary markedly in their sensitivity to off-gases.

As noted earlier, just as labeling without guidelines may be inadequate, guidelines alone without labeling also may be inadequate. Consumers may not know to adjust indoor air temperature or take other exposure-reducing actions when off-gases from pur-

---

[9]For this reason, the consumer also may not know to contact a home builder or remodeler to express concern about off-gases from cabinets or floor underlayment. If consumers *did* know to contact them, then presumably the construction industry, which constitutes a "repeat buyer" on behalf of the consumer, could respond accordingly.

[10]The Hardwood Plywood Manufacturers Association and National Particleboard Association (1991) emphasize this point in an industry update on risk assessment.

chased products deviate markedly from the standard (or when the consumer's sensitivity is higher than normal).[11]

Additional Considerations for Labeling and Standards

There are few conclusions in the economics literature as to when industry will voluntarily set standards or as to when government should intervene to set standards (for example, see discussion in Ippolito [1984]). However, there is a large literature that investigates the effect of standards on industry organization (for example, standards can drive some firms out of the market). There is also a literature that examines the effect of product liability on producer incentives to "take care," that is, to embody safety in products, provide safety information, or otherwise mitigate consumer risk (see, e.g., Assaf [1984]; Priest and Klein [1984]; Polinsky and Rubinfeld [1988]; Viscusi [1991]). Extensions of this literature suggest that producers will indeed tend to undersupply safety if they are risk neutral or if their liability can be limited by use of bankruptcy proceedings (see, e.g., Shavell [1984], Farber [1990], Ashford and Stone [1991]). Other extensions of the literature address the combined effects of product liability, liability insurance, and regulation in the form of product safety or other standards. In these cases, insurance can cause firms to undersupply safety, since insurance reduces the expected cost of litigation. However, regulation by setting standards can cause an oversupply of care (see, e.g., Shavell [1984], Farber [1990]), if firms exceed the standard (that is, by providing a safer-than-mandated product) to avoid a lawsuit (thus standards can increase a firm's expected cost of liability).

Figures 3-6 and 3-7 suggest the effects of liability, insurance, and standards on product markets. In figure 3-6, Q^* is the socially optimal output level of a formaldehyde product, represented by the intersection of supply (MPC) with a fully-informed-consumer demand curve (denoted by the solid demand curve). The industry supply curve inclusive of the expected cost of liability, $EMPC(L)$, intersects the *uninformed*-consumer demand curve to the right of Q^*, resulting in insufficient care being taken by industry. The availability of liability insurance, $EMPC(L,I)$, compounds the undersupply of

---

[11]An additional topic outside the scope of this study is worker ability and willingness to respond to workplace risk, where constraints on mobility, in particular, may loom large. For some evidence on worker responses, see Viscusi and O'Connor (1984).

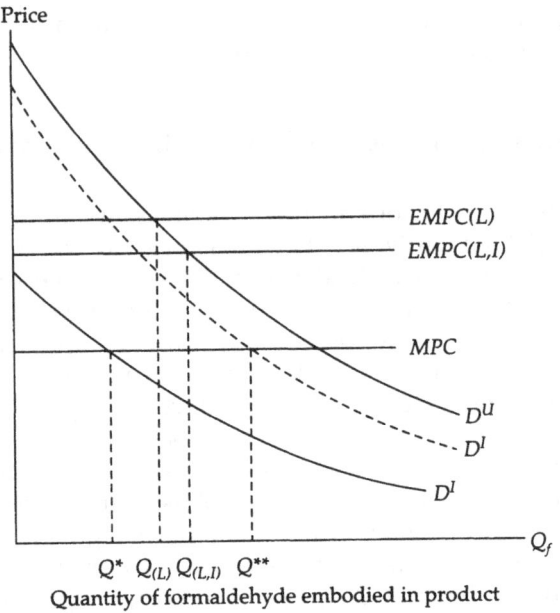

Figure 3-6. **Possible effect of liability and liability insurance on the supply of "care."**

care. By contrast, if the informed-consumer demand curve is farther to the right (denoted by the dashed demand curve) and the socially desirable output is Q^{**}, care is oversupplied. But note that in this case, information buys very little—the informed and uninformed demand curves are close together.

Figure 3-7 illustrates a setting for the oversupply of care in the presence of a product standard set at Q^*, reflecting a world characterized by perfect information on the part of the regulator (who has set the standard at the intersection of MPC and where he or she knows where the informed-consumer demand curve would be if consumers had information). In this case, the expected cost of liability is increased in reaction to a product standard and results in the oversupply of care at $Q_{(L,S)}$. An oversupply is not necessarily welfare improving—note the reduction in the consumer surplus triangle at $Q_{(L,S)}$. This result can occur when producers are sufficiently risk averse, or if they have much at stake in the event of liability-related bankruptcy proceedings.

These figures illustrate some of the difficulties of predicting the effects of product labeling and standard setting in the presence of

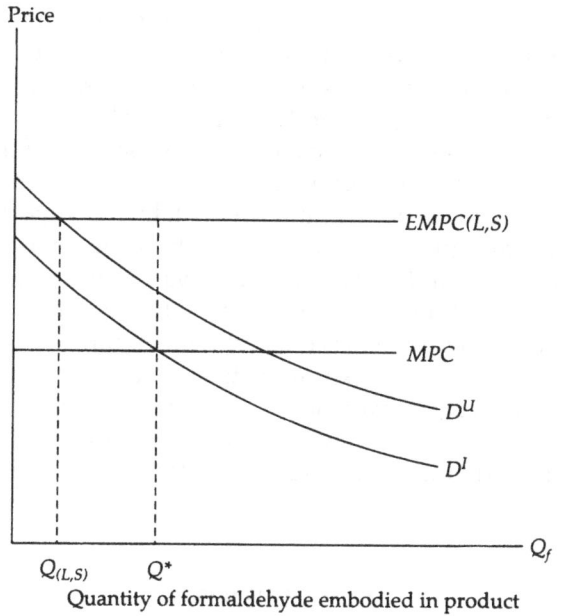

Figure 3-7. Potential for oversupply of "care" under a product standard.

product liability law and given the availability of liability insurance. Additional research to untangle these effects in the case of the formaldehyde industry would include study of risk preferences and firm capitalization structure (for example, the equity holders in the event of bankruptcy).[12]

Alternatives to Labeling or Standards

Given the difficulty of predicting market responses to labels or standards, an excise tax on final formaldehyde products might also be considered. The tax, if sufficiently high in relation to demand elasticities for these products, could operate to better ensure that consumers reduce their use of the products. Viscusi (1990) estimates

---

[12]The major formaldehyde producers appear to be highly integrated in the production of a wide variety of chemicals, and are fairly large, publicly owned companies. These factors may increase the willingness of primary chemical producers to diversify risk across different products, with some products supplied with more risk than others. Intermediate producers of formaldehyde products appear to be more specialized. If their product lines are less diverse, these producers may be less willing to take risk.

that excise taxes on cigarettes cause a decrease in consumption that approximates or exceeds levels that would occur if consumers accurately assessed the risk of smoking. Thus, taxes that increase the price of products can significantly influence behavior.

There are several disadvantages to an excise tax on formaldehyde products, however. First, product off-gases and user sensitivity are likely to vary widely; thus, a uniform tax falls disproportionately on all consumers and products and can generate a large excess burden. As noted earlier, product-specific taxes varied to reflect off-gasing would be administratively complicated (although less complicated than emission taxes or emission permits). Second, a tax does not encourage consumers to take mitigating actions such as opening a window or applying a varnish or veneer or other source barrier. Consequently, regulators may want to explore the effect of labeling and standardized guidelines before considering an excise tax on final products.

CONCLUSIONS

The discussion of incentive mechanisms for regulating production and use of formaldehyde highlights the extent of third-party effects as a determinant of intervention strategies. Third-party effects arise principally in the case of air and water pollution upstream in the formaldehyde production life cycle. However, at the final-product stage, third-party effects from formaldehyde off-gases are quite limited. The structure of the formaldehyde market (the number of primary and intermediate producers, the number of consumers, and consumer-demand characteristics such as variation in consumer sensitivity to off-gases) is also an important consideration in targeting regulation in a manner that maintains incentive effects while minimizing administrative costs.

To mitigate third-party effects that arise upstream, a product tax or tradable-emission permit based on damages and aimed at the relatively few primary producers of formaldehyde is likely to be administratively easier than a tax on intermediate producers. The tax or permit mechanisms might be regionally differentiated, with higher taxes or fewer permits in pollution nonattainment areas. Trafficking in the transportation of high-taxed (or high-permit-fee) formaldehyde to low-taxed (low-permit-fee) regions would be limited by the high transportation costs of formaldehyde.

Because of limited third-party effects associated with the use of formaldehyde-containing products, consumer information in the form of labeling and the setting of guidelines for off-gas standards would protect the more sensitive consumers, but would not unduly restrict the choices of less sensitive consumers. Such information might fill the gap created by the absence of repeat buying for many formaldehyde products (for example, furniture, cabinets, floor underlayment). It might also enable consumers suffering from irritation to distinguish formaldehyde from other possible sources of indoor air pollution as the cause. (A product's reputation, as generated through repeat buying of products or by the linking of health effects to products, typically can modify producer behavior without extra information.) However, given existing product liability law and the availability of product liability insurance, the combined effect of labels and standards is not clear. There is some evidence that information can trigger meritless litigation, with social costs that exceed the benefits of labeling or standards. It is also unclear how liability laws and insurance interact in governing producer responses to mandated standards. Producers may oversupply or undersupply safety (or oversupply or undersupply the risky attributes of the product). Anticipating producer behavior is the key to effective labeling or standards that achieve the socially desirable level of safety. Thus, the setting of broad guidelines for both labeling and standards, rather than mandating standards, is likely to be most effective in an initial approach to formaldehyde-product regulation; standards can be tightened later if experience suggests that safety is undersupplied.

REFERENCES

Ashford, Nicholas A., and Robert F. Stone. 1991. "Liability, Innovation, and Safety in the Chemical Industry." Pp. 327–367 in Peter W. Huber and Robert E. Litan, eds., *The Liability Maze: The Impact of Liability Law on Safety and Innovation* (Washington, DC: Brookings Institution).

Assaf, George B. 1984. "The Shape of Reaction Functions and the Efficiency of Liability Rules: A Correction." *Journal of Legal Studies*, vol. 13 (January), pp. 101–111.

Bower, John. 1989. *The Healthy House: How to Buy One, How to Cure a "Sick" One, How to Build One* (New York, NY: Carol Communications).

Calfee, John E. 1986. "The Ghost of Cigarette Advertising Past." *Regulation*, vol. 10, no. 2, pp. 35–45.

Costanza, Robert, and Charles Perrings. 1990. "A Flexible Assurance Bonding System for Improved Environmental Management." *Ecological Economics*, vol. 2, no. 1 (April), pp. 57–75.

Council on Scientific Affairs. 1989. "Council Report: Formaldehyde." *Journal of the American Medical Association*, vol. 261, no. 8 (Feb. 24), pp. 1183–1187.

Farber, Stephen. 1990. "Regulatory Schemes and Self-Protective Environmental Risk Control: A Comparison of Insurance, Liability, and Deposit/Refund Systems" (Baton Rouge: Louisiana State University, Department of Economics).

Fishbein, Lawrence, and Carol J. Henry. 1990. "Health Effects of Methanol: An Overview." Pp. 241–250 in Wilfrid L. Kohl, ed., *Methanol as an Alternative Fuel Choice: An Assessment* (Washington, DC: Johns Hopkins Foreign Policy Institute, Paul H. Nitze School of Advanced International Studies).

Friess, Seymour L. 1991. "An Update: Carcinogenic Risk Assessment for Indoor Exposures to Formaldehyde Gas Emissions from Certain Consumer Products." Mimeo, April 10.

Groah, William J., and Gary D. Gramp. 1991. "An Estimate of Home Occupant Exposures to Formaldehyde Gas from Unfinished UF-Bonded Wood Panel Products Used for Cabinets, Furniture and Other Industrial Applications Using a Simplified Approach" (Reston, VA: Hardwood Plywood Manufacturers Association), April 8.

Hardwood Plywood Manufacturers Association and National Particleboard Association. 1991. "California Proposition 65: Risk Assessment for Wood Products in Furniture and Cabinets" (Reston, VA: Hardwood Plywood Manufacturers Association), April 11.

Huber, Peter W. 1988. *Liability: The Legal Revolution and Its Consequences* (New York: Basic Books).

———, and Robert E. Litan, eds. 1991. *The Liability Maze: The Impact of Liability Law on Safety and Innovation* (Washington, DC: The Brookings Institution).

Ippolito, Pauline M. 1981. "Information and the Life Cycle Consumption of Hazardous Goods." *Economic Inquiry*, vol. 19 (October), pp. 529–558.

———. 1984. "Consumer Protection Economics: A Selective Survey." Pp. 1–34 in Pauline M. Ippolito and David T. Scheffman, eds., *Empirical Approaches to Consumer Protection Economics*. Proceedings of a conference sponsored by the Bureau of Economics, Federal Trade Commission, Washington, DC, April 26–27.

———. 1990. "Bonding and Nonbonding Signals of Product Quality." *Journal of Business*, vol. 63, no. 1 (pt. 1), pp. 41–60.

———, and Richard A. Ippolito. 1984. "Measuring the Value of Life Saving From Consumer Reactions to New Information." *Journal of Public Economics*, vol. 25, pp. 53–81.

Krupnick, Alan J., Margaret A. Walls, and Michael A. Toman. 1990. "The Cost-Effectiveness and Energy Security Benefits of Methanol Vehicles." Discussion Paper QE90-25 (Washington, DC: Resources for the Future).

Machiele, Paul A. 1990. "A Health and Safety Assessment of Methanol as an Alternative Fuel." Pp. 217–240 in Wilfrid L. Kohl, ed., *Methanol as an Alternative Fuel Choice: An Assessment* (Washington, DC: Johns Hopkins Foreign Policy Institute, Paul H. Nitze School of Advanced International Studies).

Magat, Wesley A., W. Kip Viscusi, and Joel Huber. 1988. "Consumer Processing of Hazard Warning Information," *Journal of Risk and Uncertainty*, vol. 1, pp. 201–232.

Phipps, Tim T., Kristen Allen, and Julie A. Caswell. 1989. "The Political Economics of California's Proposition 65." Discussion Paper No. FAP89-07 (Washington, DC: Resources for the Future, August).

Polinsky, A. Mitchell, and Daniel L. Rubinfeld. 1988. "The Welfare Implications of Costly Litigation for the Level of Liability." *Journal of Legal Studies*, vol. 17 (January), pp. 151–164.

Posner, Richard A. 1986. *Economic Analysis of Law*, 3rd ed. (Boston: Little, Brown).

Priest, George L., and Benjamin Klein. 1984. "The Selection of Disputes for Litigation." *Journal of Legal Studies*, vol. 13 (January), pp. 1–55.

Roberts, Marc J., and Michael Spence. 1976. "Effluent Charges and Licenses Under Certainty." *Journal of Public Economics*, vol. 5, pp. 193–208.

Shavell, Steven. 1984. "A Model of the Optimal Use of Liability and Safety Regulation." *Rand Journal of Economics*, vol. 15, no. 2 (Summer), pp. 271–280.

Tirole, Jean. 1988. *The Theory of Industrial Organization* (Cambridge, MA: MIT Press).

Viscusi, W. Kip. 1990. "Do Smokers Underestimate Risks?" *Journal of Political Economy*, vol. 98, no. 6 (December), pp. 1253–1269.

———. 1991. "Product and Occupational Liability." *Journal of Economic Perspectives*, vol. 5, no. 3, pp. 71–92.

———, and Joni Hersch. 1990. "The Market Response to Product Safety Litigation." *Journal of Regulatory Economics*, vol. 2, pp. 215–230.

———, and Charles J. O'Connor. 1984. "Adaptive Responses to Chemical Labeling: Are Workers Bayesian Decision Makers?" *American Economic Review*, vol. 74, no. 5 (December), pp. 942–956.

"Warning: Litigation Can Be Hazardous." 1990. *National Journal* (October 6), p. 2403.

Weitzman, Martin L. 1974. "Prices vs. Quantities." *Review of Economic Studies*, vol. 41, no. 4 (October), pp. 477–491.

4

Cadmium

Cadmium is a silver-white, soft, malleable, toxic, metallic element. Although it is widely distributed in trace amounts throughout the natural environment, the metal is infrequently found in high concentrations.

An almost unique combination of properties renders cadmium useful in a wide variety of consumer and industrial products, but, in contrast to chlorinated solvents and formaldehyde, the normal use of cadmium-containing products by consumers is rarely a source of exposure. Emissions and wastes from smelting operations and production processes and from the disposal of final products containing refined cadmium all contribute to increased concentrations of cadmium in air, water, and soil.

Because the quantities of cadmium embodied in many products are so small, cadmium recovery from these products is impractical. Among the major uses of refined cadmium, only nickel–cadmium batteries contain a sufficient proportion of cadmium to make recovery feasible. The cadmium from few batteries is recovered at present, however; instead, the batteries end up in landfills or as incinerator ash. Among other incidental contributors of cadmium to the environment are agricultural fertilizers, in which cadmium is a common associate of phosphates; dusts from cement manufacture; and emissions from coal and oil combustion.

Among the various sources of exposure to cadmium, the disposal of final products containing refined cadmium appears to be the best candidate for incentive-based approaches to control. One reason is that there are broad third-party effects associated with disposal, arising because consumers receive inadequate signals about the social costs associated with disposal. Another reason is that the variety of cadmium-containing products and the varying degrees to which suitable substitutes are available could make typical command-and-control regulatory approaches complex and costly. Accordingly, this chapter focuses on economic incentive schemes to control the use and disposal of cadmium-containing products. We consider incentive schemes that rely on taxes, tradable permits, and deposit-refund systems.

Although we focus on the use and disposal of final products, the nature of the cadmium market does require that attention be paid to the production life cycle for refined cadmium. Cadmium metal is primarily a by-product of the smelting of zinc ores. New cadmium is supplied approximately in fixed proportion to the amount of zinc produced. Therefore, efforts to reduce the demand for primary cadmium, whether through tax incentives to reduce product use or through incentives to increase recycling, may result in little or no decline in overall use (a socially efficient outcome if the tax, say, were set at the correct risk-based level). Another possibility is that a tax might lead to an accumulation of cadmium stocks at smelting plants, just shifting the location of the storage or disposal problem.

This chapter is organized as follows: The next sections discuss the production and uses of cadmium, the potential health effects and pathways of exposure during product life cycles, and the existing legislation on cadmium. Turning then to consideration of incentive-based regulatory approaches, we first emphasize some of the key attributes of cadmium markets and cadmium uses that influence the choice of a regulatory approach. Then we describe particular incentive-based mechanisms and comment on their strengths and shortcomings as compared with alternative regulatory approaches.

PRODUCTION AND USE

World Production

Primary cadmium metal production is a by-product of the smelting of zinc ores. On average, the ores contain about 3 kilograms (kg) of cadmium per 1,000 kg of zinc (Cadmium Association, 1980). To a much lesser extent, cadmium is also found in lead and copper ores. Worldwide production from the traditionally market-based economies has averaged about 14,500 tonnes (metric tons) per year over the past decade, with more recent production levels of about 15,000 tonnes (Cadmium Association, 1990). Production during the early 1990s was down somewhat from the peak production level of 1988. Refined cadmium comes primarily from a few large smelter refiners

Table 4-1.   Refined Cadmium Production in Traditionally
Market-Based Economies

	Refined cadmium (tonnes)	
	1989	1990
Japan	2,700	2,492
Belgium	1,754	1,958
United States	2,106	1,912
Canada	1,570	1,463
Germany (former Federal Republic)	1,208	973
Italy	776	668
Australia	696	638
South America	1,962	1,852
Other market economies	2,613	2,788
TOTAL	15,386	14,744

Source: Data from Cadmium Association, *Cadmium 1990: A Review* (London, 1990).

in Japan, the United States, Canada, Belgium, Germany, Australia, Peru, Italy, and Mexico.[1] These major producers of primary cadmium are all producers of zinc. Table 4-1 lists recent production statistics for the traditionally market-based economies.

Cadmium is currently recycled only from large batteries and industrial scraps, resulting in an estimated 1,500 tonnes of secondary cadmium annually (Kelecom, 1989). Plants that recycle scrap and larger batteries are located in France, Japan, South Korea, Finland, and Sweden—in 1988, 100 tonnes were recovered in Sweden, 500 tonnes in France, and 600 tonnes in the Far East (Anulf, 1989). Worldwide, about 15 percent of industrial batteries are returned for recycling (Kelecom, 1989). Because of high collection and processing costs, virtually no cadmium is recovered from household batteries. Their cadmium content can be as high as 20 percent by weight, however, which is sufficient to make recycling profitable—once the batteries are collected and separated.

U.S. Production and Trade

U.S. production of refined cadmium in 1990 was at about 1,850 tonnes per year, down about 9 percent from peak production in 1988

---

[1]The traditionally centrally planned economies are estimated to produce an additional 4,000 to 5,000 tonnes annually; both China and the former Soviet Union are large producers (U.S. Bureau of Mines [1990]).

(Cadmium Association, 1990). Only four companies in the United States (located in Colorado, Illinois, Oklahoma, and Tennessee) produce cadmium. There are no major recycling plants in the United States.

Less than half of the U.S. demand for refined cadmium is met by domestic output. Total U.S. consumption of refined cadmium has averaged about 4,000 tonnes annually, with the difference being made up by imports. The United States imports more refined cadmium from Canada than from any other country (41 percent of imports). Other major sources are Australia (17 percent), Mexico (16 percent), and Germany (11 percent) (U.S. Bureau of Mines, 1990).

Additional cadmium enters the country embodied in final products. Nickel–cadmium batteries, the primary use of cadmium, are produced by six firms in the United States. These firms now supply only about half of the batteries sold in this country. The other half are imported, with Japan the largest foreign supplier. Numerous domestic and foreign firms also use cadmium as a stabilizer for plastics.

The Market for Cadmium

The market price of cadmium has been variable over the past 10 years, and extremely so over the past 5 years. During the early 1980s, prices gradually declined, dropping from about $3.00 per pound in early 1980 to less than $1.00 per pound by the end of 1986. Beginning in 1987 cadmium prices rose, as prices did for most metals, but spectacularly—to over $8.00 per pound by 1988. Then they dropped sharply during 1989 and, after a brief rally, continued to fall during 1990. In 1990 the price fell from a high of $5.20 per pound in January to a low of $1.15 by November. By 1991, it had recovered to about $3.00 per pound (Cadmium Association, 1990).

Cadmium prices seem to have been influenced by two primary forces. The gradual decline in prices during the early 1980s is associated with the increased concerns over the environmental hazards of this substance. The tightening of regulations on the use of cadmium, particularly in Japan and Sweden, led to some reduction in consumption and to anticipation of further declines. In addition, cadmium supply was increasing, in part because refiners found it necessary to reduce the residual cadmium content of their zinc in order to satisfy the health concerns of zinc purchasers. However, toward the end of the 1980s, growth in demand for rechargeable nickel–cadmium batteries accelerated. Users apparently then began to stockpile cadmium, driving prices up in a speculative flurry. (The inelastic supply of refined cadmium, as with other minor metals, can lead to such wild price swings.) It appears that the subsequent price drop was a normal market adjustment.

Properties of Cadmium

Cadmium has a number of unusual and desirable features that result in its use in a wide variety of products (Cadmium Association, 1978a, 1978b, 1978c, 1979, 1980). The pure metal oxidizes slowly to form a coating that both protects against corrosion and offers a high degree of natural lubrication; it is a soft metal that is easily made into sheets or drawn into wires; it is a good conductor of electricity; and it combines easily with nonferrous metals to make alloys with desirable characteristics.

The compounds of cadmium are similarly versatile. Cadmium sulfide, the form in which cadmium occurs as ore, is widely used for its optical and electrical properties. Cadmium sulfide is the basis for pigments in artists' paints and in plastics, glasses, and ceramics. It is desirable both for the bright reds, oranges, and yellows that can be attained and for the stability of these pigments under processing heat and exposure to outdoor light. The sulfide is also now used in some photovoltaic, photoconductor, and semiconductor applications. The low solubility of cadmium sulfide makes it one of the less toxic cadmium compounds,

Cadmium oxide is used as a stabilizer for polyvinyl chloride plastics, making them less susceptible to degradation under direct sunlight and high temperatures. It is also used in manufacturing the negative electrode in nickel–cadmium batteries. However, the cadmium compounds in batteries are somewhat more soluble and toxic than those in pigments.

The Uses of Cadmium

There are five major uses for cadmium and the cadmium compounds: batteries, pigments, plastic stabilizers, coatings, and alloys.[2] The rechargeable nickel–cadmium battery is now the primary use; ten years ago, the dominant use was the plating of steel for corrosion protec-

---

[2]Much of this section is based on Hiscock and Volpe (1989), Volpe (1985), and Wilson and Volpe (1983, 1986).

Table 4-2. Estimated Market Share of World Cadmium Use by Product Type

	Percentage of market share	
	1980	1990
Batteries	23	55
Pigments	27	20
Stabilizers	12	10
Coatings	34	8
Alloys	3	3
Miscellaneous	1	4

Source: Adapted, with permission, from Cadmium Association, *Cadmium 1990: A Review* (London, 1990).

tion, but this application has declined significantly. Use in pigments and stabilizers has declined somewhat since the late 1980s, although the rate of decline has slowed. Table 4-2 shows world trends in the use of cadmium in these products (Cadmium Association, 1990).

Trends in the U.S. cadmium market are similar to the world trends, with most of the growth in demand attributable to increased use of nickel–cadmium rechargeable batteries. Demand for these batteries is expected to continue to drive the market for cadmium. It is also worth noting that the U.S. government stockpiles cadmium as a strategic material, presently holding almost 3,000 tonnes in inventory. It has not been active in adding to these stocks, despite authorization to hold 5,300 tonnes (U.S. Bureau of Mines, 1990).

Batteries. Rechargeable batteries are the only use of cadmium that has been steadily growing. This growth in demand has come from the consumer electronics sector, although industrial use remains high. Large industrial applications—including wide use of nickel–cadmium batteries for emergency backup power and for starting large engines such as those in planes and locomotives—account for about one-third of the cadmium used in batteries. Small consumer applications—for example, rechargeable flashlights and laptop computers—account for the other two-thirds (Kelecom, 1989).

Nickel–cadmium batteries have many desirable features: recharging characteristics, portability, tolerance of temperature extremes, and long life (Cadmium Association, 1979). Substitutes exist for all uses, including lead–acid batteries for industrial use and non-rechargeable batteries for many consumer uses (McClure, 1989). There is also increasing interest in less toxic rechargeable batteries for consumer products, such as metal–hydride batteries. However, at current cadmium prices, demand for nickel–cadmium batteries is expected to continue to increase.

Coatings. Cadmium coatings on steel reduce corrosion and friction and provide easy solderability, characteristics that are quite useful to the electronics industry (Cadmium Association, 1978a). Bolts and connectors in the automotive, electronics, aerospace, and offshore oil industries are often cadmium-coated. Cadmium is particularly useful in preventing contact corrosion when steel must be joined to aluminum. However, the use of cadmium coatings worldwide is declining because of environmental regulation and health concerns within the plating and steel-reprocessing industries. It is likely that some cadmium plating will continue to be required in certain aerospace and marine applications where the risk of corrosion is high. However, zinc–nickel alloys are attractive substitutes in many cases.

Pigments. Cadmium pigments are used mostly in plastics. Their use in plastics accounts for 90 percent of the cadmium in pigments worldwide. Chosen for their color, weather resistance, and ability to withstand high processing heat (Cadmium Association, no date), cadmium pigments are most frequently used in plastics for outdoor applications. For plastics, these pigments are perhaps essential only in technical applications that require plastics having high processing temperatures and that for some reason need cadmium pigment coloration.

Cadmium pigments are also used in ceramics and glass. In these applications, no substitutes exist for a pure red that can withstand high processing temperatures. Substitutes for cadmium yellows do exist, but they are not suitable for all uses. Cadmium pigments are also needed for their pure colors in certain optical applications, particularly signal lenses (traffic and airport lights). Artists' paints and inks account for a very small amount (about 1 percent) of cadmium used in pigments.

Stabilizers. Barium–cadmium stabilizers are added to plastics to protect against breakdown caused by weather and light exposure (Cadmium Association, 1978c). Window frames account for about 75 percent of the cadmium used in stabilizers); such products as outdoor furniture, plastic pool liners, and roof sheeting account for the remainder. The use of cadmium as a stabilizer in polyvinyl chloride plastics was once considered almost essential for outdoor applications. However, the 1982 ban (loosely enforced) on cadmium stabilizers in Sweden initiated a slow process of substitution, especially in Europe (Kjallman, 1989). Recent advances in calcium–zinc stabilizers used in conjunction with costabilizers suggest that these may be adequate nontoxic alternatives. Certain other substitutes, such as those that contain lead or barium–zinc stabilizers, may be no more desirable than cadmium because of their potential toxicity.

Alloys. Cadmium is used in small amounts to improve strength, wear resistance, and electrical properties in copper, tin, lead, and zinc alloys (Cadmium Association, 1978b). Such alloys are found in gas turbines, nuclear reactors, and bearings; demand in these uses is relatively stable. Cadmium is also used to lower melting points in such products as solder, welding rods, and brazes.

EXPOSURE PATHWAYS AND HEALTH EFFECTS

The primary toxic effect of cadmium results from the accumulation of cadmium compounds in the kidneys. Pulmonary lung diseases may be associated with continued inhalation of cadmium. The substance is also considered a possible carcinogen, having some apparent association with lung and prostate cancers.

Cadmium is naturally present in soils at low levels; thus, small dietary intakes of cadmium are normal. Smoking may add perhaps half again as much to the average rate of cadmium intake. For nonsmokers, it is estimated that 50 years of dietary intake at high levels could lead to potentially damaging concentrations of cadmium in the kidneys for 10 percent of the population. On this basis, the Food and Agriculture Organization and the World Health Organization have suggested provisional guidelines for tolerable weekly intake. (Lauwerys and Malcolm [1985] and the Organisation for Economic Co-operation and Development [1991] provide reviews of current knowledge and literature on the health effects of cadmium.)

Cadmium enters the body either through ingestion or inhalation. Absorption through the skin is negligible. The primary health concern is with the inhalation of dust and fumes in the workplace. Of particular concern for workers is that elemental cadmium vaporizes easily when heated, creating a fine mist of cadmium oxide that can be absorbed through the lungs fairly readily. Such fumes can result from welding, plating, and furnace emissions. The degree of industrial exposure depends upon the nature of the production process. Cadmium refining results in dust generated during the crushing of ores and in stack emissions from furnaces. Production or use of cadmium stabilizers and pigments also generates dust, although much has been done to reduce that dust. In addition, pigments are among the least soluble cadmium compounds, making toxicity low. The manufacture of batteries can result in cadmium oxide fumes and dust at various stages of the process. Emissions during recycling are similar to those during initial refining, with crushing and furnace operations producing dusts and fumes.

The greatest concern for the general public appears to be the long-term accumulation of cadmium in soils where it is taken up by plants, including food crops and tobacco. Increments to background cadmium levels may come from the deposition of airborne emissions (oil and coal combustion) and from fertilizers, sewage sludges, and waste and disposal. Although cadmium in landfills does leach into groundwater, past studies (Bromley and coauthors, 1983) suggest that these levels in groundwater, on average, do not exceed the U.S. Environmental Protection Agency standard for cadmium in drinking water. This reflects the limited solubility of cadmium and the dilution of cadmium content away from disposal sites. Thus, contamination of drinking water supplies seems unlikely to be a major source of increased general exposure. Inhalation of airborne emissions is also not likely to be of great concern other than in areas close to the source of emissions. However, both air and water transport of cadmium into soils may be of concern. Direct contact with products containing cadmium is not a source of risk, unless there is accidental ingestion.

Table 4-3 summarizes the exposure points during the life cycle of cadmium products. "Incidental cadmium products" refers to products such as cements, fertilizers, and coal that contain cadmium but to which cadmium is not deliberately added. Thus, exposure could occur during the burning of coal and oil or the application of fertilizers. The potential exposure associated with disposal appears to be greater for products that are incinerated than for those sent directly to landfills, since the cadmium in incinerator ash is more concentrated. Landfill disposal of batteries seems to present a somewhat higher risk than does landfill disposal of plastics, because cadmium leaches more rapidly from batteries (Bromley and coauthors, 1983). The greatest problems are to likely to be faced by future generations, as the gradual leaching of cadmium from landfills and the continuing deposition from other sources leads to a buildup of cadmium levels in soils.

EXISTING STATUTES AND REGULATIONS

Table 4-4 lists some of the existing legislation on cadmium. Pending the outcome of interpretation of some of these rules, cadmium-

Table 4-3. Cadmium Life Cycle and Sources of Exposure

Stage of production or use	Exposure
Extraction/import	Worker
Refining	Worker
	Ambient air and water
Production	
Refined and incidental	Worker
cadmium products[a]	Ambient air and water
Distribution	None
Consumption	
Refined cadmium products	
–Household use	Consumer
–Commercial use	Worker
Incidental cadmium products	
–Household use	Ambient air, water, and soil
–Commercial use	Worker
	Ambient air and water
Disposal	
Commercial and household	Ambient air, water, and soil
Reprocessing/recycling	Worker

[a]"Incidental cadmium products" refers to products such as cements, fertilizers, and coal that contain cadmium but to which cadmium is not deliberately added.

contaminated incinerator ash may have to be treated as hazardous waste, requiring dumping in specially lined landfills at high costs. As of 1992, the Occupational Safety and Health Administration (OSHA) was considering the tightening of standards on occupational exposures, with proposed new limits on exposure to dust and fumes much below current standards. Industry groups claim that the proposed new standards, along with the siting and production costs of meeting existing environmental regulations, are dampening interest in building battery recycling plants. It is also likely that the collection of nickel–cadmium batteries and their transport for recycling would be impeded by regulations on the storage and transport of hazardous wastes. State interest in regulating cadmium in batteries is also increasing. In 1990, Minnesota and Connecticut passed legislation that is intended to eventually remove nickel–cadmium batteries from the waste stream (see Erickson [1991]).

In addition to U.S. regulation, controls on cadmium in Europe and Japan have had a strong worldwide effect in reducing the demand for cadmium (Kjallman, 1989; Shagarofsky-Tummers, 1989). In 1979 Sweden banned the use of cadmium in coatings, pigments, and stabilizers. Many limited exemptions were offered, allowing indus-

Table 4-4. Statutes and Regulations Pertaining to Cadmium

Responsible agency	Statute or regulation	Uses or sources of exposure covered
Occupational Safety and Health Administration		Limits workplace exposure to 200 micrograms per cubic meter ($\mu g/m^3$) for cadmium dust and 100 $\mu g/m^3$ for cadmium fumes (time-weighted averages).
U.S. Environmental Protection Agency (EPA)		Sets maximum contaminant level for drinking water and ambient water quality of 10 $\mu g/liter$; for levels in solid waste applied to farmland, of 0.5 kilograms per hectare (kg/ha) (annual rate).
EPA	Clean Air Act	Regulates cadmium as an air toxic.
EPA	Comprehensive Environmental Response, Compensation, and Liability Act	Regulates cadmium as a hazardous waste.
EPA	Resource Conservation and Recovery Act	Regulates cadmium as a hazardous waste.

tries time to address the technical problems of reduced cadmium usage. Nonetheless, experts estimate that cadmium usage was reduced by 90 percent over a 10-year period. In other European countries, the demand for cadmium has also declined because producers have found it necessary to reduce use in order to maintain their access to the Swedish market.

KEY ATTRIBUTES OF REGULATORY INTEREST

This section summarizes characteristics of cadmium and cadmium markets that may influence forms of incentive-based regulation. Perhaps foremost among those are the very small amounts of cadmium embodied in many products, the wide variety of cadmium-containing products, and the considerable variation in the degree to which substitutes for cadmium are available. In addition, third-party exposure is related primarily to the disposal of cadmium products rather than to direct product use. Among products there is some variation in the potential for exposure; disposal of the more soluble (and thus more toxic) compounds in batteries may be associated with greater exposure potential than disposal of the pigments and stabilizers in plastics. Across disposal practices, the risk of exposure is generally greater from incineration than from direct landfill disposal.

The variety of products containing cadmium reflects the desirable and unique properties of this substance. For certain applications, substitutes for cadmium are not attractive in terms of cost or suitability; therefore, across-the-board limits on cadmium use might result in considerable forgone consumer benefits (through increased product costs). The association of third-party damages with disposal rather than use suggests that regulation should, if possible, be designed to provide incentives that redirect disposal so as to reduce these damages. However, the quantity of cadmium in many products is so small that controlled disposal or recycling may be impractical or costly (with the nickel–cadmium battery perhaps an exception). For example, it is virtually infeasible to recover the very small amounts of cadmium in plastics or in corrosion-proof coatings. Managing disposal—for example, by requiring that cadmium-containing products be separated in the waste stream—is difficult because the products are frequently large and bulky; ascertaining if they contain cadmium is difficult; and collection, separation, and disposal or storage could be costly.

There are few producers or importers of refined cadmium, many producers and importers of cadmium-containing products, and a very large number of household products containing cadmium. With the bulk of refined cadmium going to consumer products, cadmium disposal is largely in the form of household waste.

Each of these factors suggests that upstream regulation of refined cadmium might be administratively easier than the regulation of its use in products or the regulation of product disposal. However, such upstream regulation could be unnecessarily costly to consumers if it restricted use in general rather than targeting disposal choices or particular uses that are the sources of greatest third-party damages.

Also of importance are supply conditions for cadmium. The worldwide supply of refined cadmium is essentially fixed in proportion to the production of zinc. Thus, even if cadmium use were to be phased out, some supply would continue to be available to the extent that zinc production is unchanged by efforts to control cadmium. Stocks of unsold cadmium or partially processed cadmium ores could thus accumulate at a few sites. Without proper management to control soil contamination or other environmental harm, these sites might pose greater environmental risks than the more dispersed (but less concentrated) disposal of the cadmium that remained embodied in products. Nonetheless, because there are few refiners and a relatively small tonnage of cadmium produced annually, it may well be less costly to deal with unsold cadmium at the refinery than with the disposal of final products. Of course, for the United States, a considerable reduction in use could be achieved through reduced imports, moving much of the potential problem with unsold cadmium offshore.

Because the world supply of refined cadmium is likely to be fairly inelastic, it may not make economic sense to encourage recycling of cadmium from household batteries (particularly after the possible environmental costs associated with recycling are considered). Recovered cadmium would supplement, rather than replace, the production of newly refined cadmium. Under these circumstances, policies designed to artificially increase the recycling of cadmium might drive down cadmium prices, perhaps sharply, reducing returns in the recycling industry. This effect could make the policy costly to sustain and require subsidies to the recycling industry.

As noted earlier, nickel–cadmium batteries may be an easier target for recycling. At present, most of these batteries are discarded. However, they do appear to contain enough cadmium to make recovery practical once the batteries are collected and sorted. Further, these batteries are sufficiently compact that controlled disposal or even storage might be feasible, although perhaps costly.

Several factors complicate the collection of nickel–cadmium batteries (or mechanical separation of these batteries from the waste stream). First, removal of nickel–cadmium batteries from consumer electronic items can be difficult, because many of these batteries are integrated into the product (for example, in rechargeable flashlights).[3] Either the products would have to be redesigned to allow battery removal or the entire product would have to be accepted for return. Adequate labeling would help to inform consumers of the presence of the nickel–cadmium battery and the return policy.

Second, households discard few nickel–cadmium batteries and, in the absence of labeling, it may be difficult to distinguish them from other batteries. It therefore seems unlikely that large quantities of nickel–cadmium household batteries can be collected separately from other battery types.[4] With currently practical recycling tech-

---

[3]The batteries are built in partly for safety reasons, since replacement with the wrong battery type can lead to fire or explosion.

[4]Note also that with cadmium classified a hazardous waste, handling of collected

nologies, such separation is necessary. However, separation would not be necessary if the goal of collection were simply the controlled disposal or storage of all batteries. In this case, household batteries could be removed from the waste stream by returning them to the retailer or manufacturer, through household pickup programs, or by mechanically separating them from wastes at the disposal site.

Whether practical disposal options exist for batteries after collection is an empirical question. It might prove useful simply to ensure that the collected batteries are not incinerated, but are dispersed to landfills, or even to specialized landfills or disposal sites. As noted earlier, although the practicality of recovering metals from unsorted batteries is unclear, it is at least theoretically possible that subsidized recycling would prove cheaper than controlled disposal or storage.

In general, with some disposal costs an unavoidable result of battery use, it is appropriate that there should be incentives that limit use as well as redirect disposal (to the extent that options are available). In the absence of practical collection and disposal options, batteries should be treated the same as other cadmium-containing products, with incentives that encourage reduction in use.

PROPOSED INCENTIVE MECHANISMS

In the case of cadmium, unregulated third-party effects are related to consumer disposal of cadmium products. Disposal, particularly through incineration, leads to a gradual accumulation of the toxic metal in soils, with an associated increase in public exposure and health risks. Although it is not clear whether these expected damages are sufficient to justify regulation, for our purposes we take it as given that cadmium is to be regulated. From our perspective, then, the problem is to provide adequate market signals to encourage cost-effective use and disposal of cadmium. Because of differences in the nature of cadmium products, we suggest two separate mechanisms to internalize the third-party effects of disposal.[5]

Accordingly, table 4-5 outlines our discussion of regulatory options for coatings, pigments, and stabilizers, and table 4-6 does the

nickel–cadmium batteries could require costly permit schemes or other practices unless exemptions are arranged.

[5]In each case, it should be remembered that costs associated with the accumulation of unsold or unprocessed cadmium at the refinery may offset gains from limiting the disposal of cadmium.

Table 4-5. Regulatory Options for Cadmium Coatings, Pigments, and Stabilizers

Life-cycle stage	Regulatory option	Verdict	Comments
Manufacture	Content labeling	Maybe	Provides limited options for changing response.
	Safety labeling	No	Is not relevant, given no private risks to user.
	Ban or limit on cadmium use	No	Imposes loss of user benefits, industry cost.
	Tradable cadmium permits	Yes/maybe	Reduces use and disposal externalities; reduces cost relative to direct ban; may have high administrative costs; may be difficult to administer for imports containing cadmium.
	Cadmium input tax	Yes/maybe	Reduces use and disposal externalities; may be ineffective if cadmium supply is inelastic.
Product use	Product sales tax	No	May be easier to tax cadmium input.
	Safety labeling	No	Is not relevant, given no private risks to user.
Product disposal	Disposal fees	No	May be impractical to target cadmium products; may offer little opportunity to alter disposal practice; has some incentive for illegal disposal.
	Deposit-refund schemes	No	May offer little opportunity to alter disposal practice.
	Tradable permits for recycled content	No	May be impractical; costs of recycling may be high, with cadmium supply fixed to zinc production.

Table 4-6. Regulatory Options for Cadmium Batteries

Life-cycle stage	Regulatory option	Verdict	Comments
Manufacture	Content labeling	Maybe	May be useful if discarded batteries must be sorted.
	Safety labeling	No	Is not relevant, given no private risks to user.
	Require removable battery	Maybe	Improves collection efficiency; adds some consumer safety risk.
	Ban or limit cadmium	No	Imposes loss of consumer benefits, industry cost.
	Tradable cadmium permits	No	Reduces use and related disposal externalities; does not alter disposal incentives; limits in growing market may be difficult to set; may have high administrative costs.
	Cadmium input tax (with no refund scheme)	No	Does not alter disposal incentives, although it does reduce use.
	Cadmium input tax (with refund scheme)	Yes	As part of deposit-refund scheme: discourages cadmium use, encourages substitutes, provides money for refund on collected batteries.
Use	Sales tax	No	May be easier to tax cadmium inputs.
Disposal	Disposal fees	No	May be impractical to target the cadmium battery; gives incentive for illegal or hidden disposal.
	Subsidize collection	Yes	As part of deposit-refund scheme: provides incentives for removing cadmium from normal waste stream; reduces disposal externalities.
	Tradable recycled content permits	No	May be impractical and have high costs, with cadmium supplies fixed to zinc production.

same for batteries. (We omit alloys here as they are a small percentage of the market; see table 4-2.)

Cadmium coatings, pigments, and stabilizers generally represent unrecoverably small amounts of cadmium. For these uses we suggest either a tax on cadmium use or a phasedown of cadmium use under a tradable-permit scheme. Batteries contain relatively large amounts of cadmium in a compact and potentially recoverable form. For nickel–cadmium batteries, despite our somewhat pessimistic comments on the availability of disposal options, we suggest a variant of a deposit-refund system. The refund mechanism could be designed to encourage a search for the least-cost means of collection, which might not necessarily mean a return of batteries to a retailer or manufacturer as in a typical deposit-refund program.[6]

Incentives to Reduce Use in Coatings, Pigments, and Stabilizers

Taxes and tradable permits are two approaches that could reduce uses of cadmium in coatings, pigments, and stabilizers. It is uncertain how a tax would affect cadmium use, given an essentially fixed worldwide supply of refined cadmium. If the tax were set at the correct risk level, however, then a socially optimal outcome would be conceivable even if there were little reduction in the use of cadmium. A tradable-permit system to phase down cadmium use provides greater certainty that use of cadmium would be limited.

The appropriateness of either policy depends on the implicit assumption that any reduction in the use of cadmium is equally beneficial, independent of the specific product use or the location of disposal. More complex policies would be needed in the case of locational hot spots, or if particular high-risk products needed to be targeted separately. Of course, neither approach would directly influence disposal decisions.

A Cadmium Tax. As one option for the regulation of cadmium use in coatings, pigments, and stabilizers, we suggest a tax on purchases of refined cadmium, imported cadmium additives, and imported products containing cadmium. Such a tax, if passed through in higher product prices, would reduce the demand for cadmium, encouraging a shift to substitutes.

---

[6]Erickson (1991) describes some of the existing state legislation aimed at removing cadmium batteries from the waste stream. State programs currently combine deposit-refund, labeling, and mandated rates of battery collection.

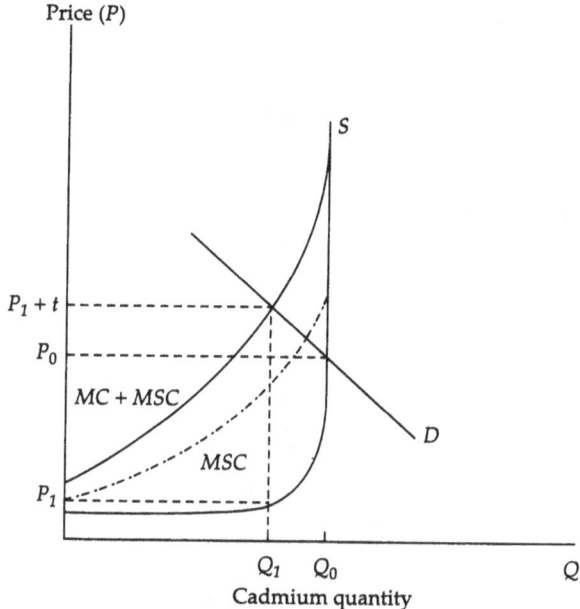

Figure 4-1. Effect of a tax on refined cadmium supplies.

Figure 4-1 illustrates the principles behind an incentive-based tax policy. We treat the U.S. market here as if it were isolated from the rest of the world. A demand curve representing the marginal benefits from cadmium use is shown by the line D. The supply curve for newly refined cadmium is labeled S. This supply curve is vertical at higher prices, with cadmium supplied in fixed proportion to zinc production. At lower prices the curve may have a flatter slope, as refiners begin to find that it does not pay to fully process some of the separated cadmium ores. In this lower segment, the supply curve represents the marginal costs of cadmium production.

The marginal social costs of cadmium disposal (MSC) are drawn to increase with higher rates of usage. The curve $MC + MSC$ shows the full marginal cost of cadmium production, summing the private marginal costs of production and the social costs of product disposal.

In the absence of a tax, the private market leads to a price P_0 with consumption Q_0. At this level of consumption, marginal benefits (P_0) are considerably less than the full marginal costs (from the curve $MC + MSC$). Net benefits can be increased by reducing consumption to Q_1, at which point marginal benefits just equal the full marginal costs. A cadmium tax t, set to equal MSC at Q_1, would serve to reduce equilibrium consumption to this level Q_1. With the tax, the

97

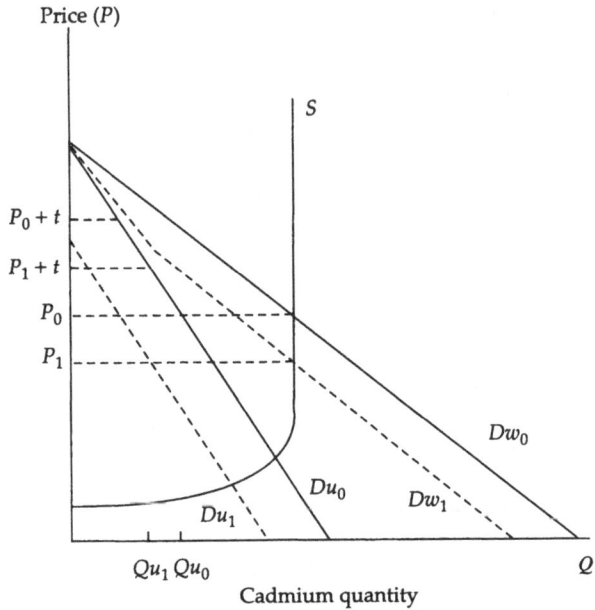

Figure 4-2. **Effect of a U.S. tax on cadmium prices and consumption.**

refiners offer Q_1 for sale at price P_1, and consumers (that is, the consuming industry) demand Q_1 at the after-tax price of $P_1 + t$.

Figure 4-2 illustrates the world market for newly refined cadmium. Of some concern is that with a fixed worldwide supply of refined cadmium, and with demand-reducing efforts likely to be matched in other major consuming countries, a domestic tax may not be passed on in higher prices. With an inelastic supply of cadmium, much of the burden of a tax could fall on the refiners, with perhaps only modest effects on production levels or final prices. In these circumstances, a tax would raise revenue, but might have only a limited effect on cadmium use. If the tax rate is optimally set, however, then the socially desirable outcome is at least conceivable.

In figure 4-2, the world supply curve is represented by the curve S. Current world demand is represented by Dw_0, while current U.S. demand is represented by Du_0. After a tax t is imposed on U.S. consumption, effective U.S. demand as a function of the world price can be shown by the dashed line Du_1—a shift down in demand curve Du_0 by the amount of tax t. (Demand at the after-tax U.S. price is unchanged and is still given by the curve Du_0). The shift in U.S. demand leads to a corresponding leftward shift in overall world demand to curve Dw_1.

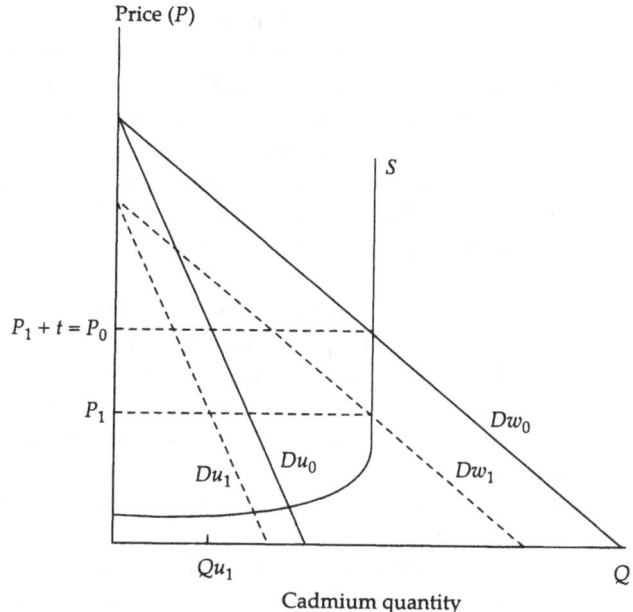

Figure 4-3. Effect of a U.S. cadmium tax matched by a world tax.

As drawn in figure 4-2, the effect of the U.S. tax is to lower the world price of cadmium from P_0 to P_1 and raise the after-tax U.S. price to $P_1 + t$. Worldwide consumption remains unchanged (reflecting inelastic world supply), but U.S. consumption declines from Qu_0 to Qu_1. Although U.S. consumption declines, this decline is somewhat less than that which would result if world cadmium prices had remained unchanged (with an after-tax price of $P_0 + t$). The smaller the U.S. share of the world market, the more effective might be a U.S. tax policy in reducing U.S. consumption (that is, an offsetting decline in world cadmium prices would be less). With current U.S. consumption representing only about one-quarter of free-market cadmium supplies, and with half of U.S. needs met by imports, it is reasonable to expect that a domestic tax would serve to raise U.S. prices and to reduce domestic cadmium use.

Figure 4-3 illustrates the effect of demand-reducing measures that might also be instituted in other major consuming countries. The specific nature of these controls will affect the world allocation of cadmium usage. In figure 4-3, the U.S. tax is assumed to be matched by an identical tax in other consuming countries. The effective world demand curve now also shifts down by the amount of the tax. In the case illustrated, the world price of cadmium declines to exactly offset the tax. The cost of purchasing cadmium is unchanged, and U.S. and world consumption is unchanged. Only at higher tax levels, sufficient to shift the effective world demand curve down to cross the supply curve in its more horizontal section, would cadmium production and consumption be reduced (but with unprocessed cadmium accumulating at smelting sites). Note that if in fact any reduction in world cadmium use is justified (under the principles illustrated in figure 4-1), such a high tax level would be appropriate.

A few points should be mentioned. Because little of the tax is likely to be passed through in the price of cadmium, most of it would probably be passed on through the price of zinc. If the supply of zinc is relatively elastic (which is likely because there are large deposits of ores worldwide), then zinc consumers would bear much of the brunt of the cadmium tax. Thus, a tax on cadmium would simply cause more of the joint production costs of cadmium and zinc to be reallocated to zinc output.

As noted earlier, a cadmium tax set at the correct level (where full marginal cost equals demand) is conceptually appropriate even if prices of both zinc and cadmium rise only slightly and consumption changes very little. If tax revenues are not used to reduce risk associated with cadmium, however, the socially desirable level of risk reduction associated with cadmium may be forthcoming only if taxes are set high enough to decrease demand significantly.[7]

Such a high tax may be perceived as punitive and therefore as politically unacceptable. Further, if our assessment of supply is correct, small misjudgments as to appropriate tax could lead to dramatic swings in cadmium output. On the other hand, enforcement of a cadmium tax should be relatively easy, since there is only a small number of refiners and importers of refined cadmium. The primary enforcement difficulty would be in assessing the cadmium content of imported products. Finally, although the tax described here does not apply to cadmium for nickel–cadmium batteries, it would be administratively advantageous if the same tax were applied to all cadmium. For batteries, that tax could serve as the "deposit" in the suggested deposit-refund system.

---

[7]More precisely, alternative uses of tax revenues could improve welfare if, for example, the revenues permitted a reduction in the rate of distorting taxes in other sectors of the economy. At issue is the linking of benefits and costs of the intervention. For additional discussion of tax changes in a general-equilibrium setting, see Oates (1991) and Goulder (1991).

A Permit-Trading Phasedown. A permit-trading scheme for the gradual phasedown of the use of cadmium is an alternative means to regulate the uses of cadmium in coatings, pigments, and stabilizers.

As a practical matter, the cadmium market seems to be suited to such a permit-trading system. First, it may be administratively easier to target directly the quantity of cadmium rather than its price, particularly given the doubts as to whether a cadmium tax will reduce use. Second, the use of cadmium, other than for nickel–cadmium batteries, is already stable or declining. Quantity targets using permits might therefore be more easily agreed upon than in a market with growing demand (that is, pressures to release the quantity cap may be less in a declining market). Finally, because of the small number of refiners and importers of refined cadmium, policing the purchases of cadmium metal may not be too difficult. Control may also be relatively easy for imports of intermediate products (such as pigments and stabilizers) that contain cadmium. Policing would be most difficult for final products that embody these intermediates (for example, plastics that may contain cadmium pigments and stabilizers). One troublesome problem might be the loophole created if purchasers of refined cadmium for nickel–cadmium batteries were exempt from the permit program. These purchasers might resell cadmium if the taxes paid on cadmium for batteries were less than the cost of purchasing permits.

The tradable-permit scheme is best viewed in comparison to the likely ban or cutback on the uses of cadmium that might be expected under command-and-control regulation. In some applications, cadmium remains strongly preferred over possible substitutes. A general ban or proportional cutback in cadmium use would impose high costs on these uses, but a tradable-permit scheme, like a tax mechanism, would allow market forces to allocate cadmium to these uses where it is of high value. In addition, the process of reallocation through taxes or trading avoids a burdensome administrative process of determining exemptions from the ban, such as occurred under Sweden's program to restrict cadmium.

A tradable-permit mechanism for reducing the use of cadmium would begin by allocating annual permits to existing purchasers of refined cadmium and imported intermediate cadmium products. The initial permits would authorize the use of cadmium at levels somewhat below existing use. Each year, the amount of cadmium permitted would be reduced. These permits would be tradable among firms. The program would be terminated after some interval (perhaps 10 years), at which point cadmium use would be either eliminated

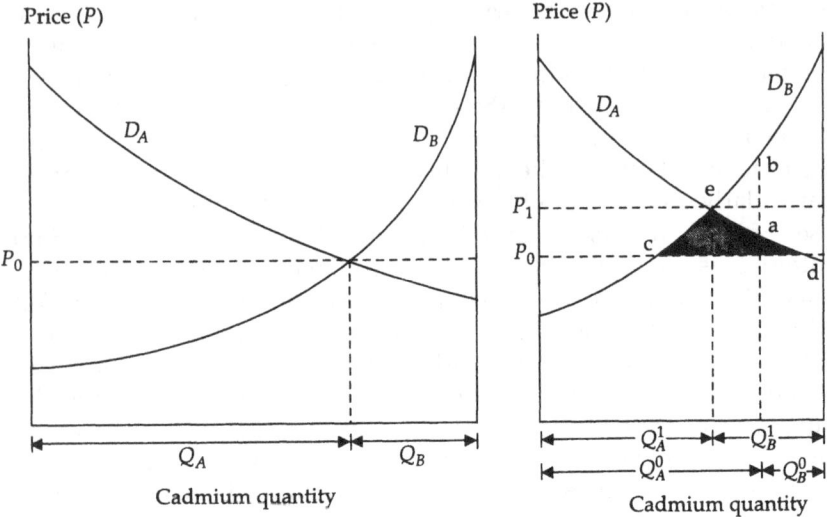

Figure 4-4. **Effect of permit trading to phase down cadmium use.**

or limited to a rate of use determined for each product type. The procedures could be essentially the same as those used in the successful phasedown of lead in gasoline (Hahn and Noll, 1990).

During the phasedown, industries that could easily substitute away from cadmium use would do so, selling their permits to firms that found it more costly to reduce cadmium use. Overall use of cadmium would be cut back to the administratively determined level, with the reduction in use allocated among firms in a cost-efficient manner as a result of the trades.

Figure 4-4 illustrates a permit-trading mechanism for cadmium. The left panel of the figure shows the market allocation of cadmium between two firms prior to any restriction on use. The marginal value of cadmium to firm A is represented by the input demand curve D_A. Similarly, the marginal value of cadmium to firm B is represented by the demand curve D_B. This second demand curve is drawn with respect to the right-hand vertical axes, in a reverse of the usual manner. At the market price of cadmium, P_0, industry A purchases quantity Q_A while industry B purchases quantity Q_B, with the marginal value of cadmium equal for each firm. Total cadmium purchases are represented by the length of the horizontal axis. The steeper slope of firm B's demand curve at the current equilibrium indicates that this firm finds it more costly to substitute away from cadmium.

Now suppose that each firm is given permits allowing the use of just 70 percent of current cadmium purchases. In the right panel, the same industry demand curves have been redrawn. The length of the horizontal axis is now reduced, indicating the now-restricted amount of cadmium permitted for use. The initial allocation of permits between firms is given by the amounts Q_A^0 and Q_B^0. At this allocation, the marginal value of cadmium to firm B exceeds the value to firm A (compare point b to point a on the demand curves). Permits would be supplied by firm A to firm B until the final allocation of cadmium use shifts to Q_A^1 and Q_B^1, with permits moving to the firm that finds it more costly to reduce its use of cadmium. At the equilibrium allocation, cadmium is of equal marginal value P_1 to each firm. Cadmium itself would sell for the price P_0 (or less, depending upon the extent to which reduced consumption leads to lower world prices) and the permits would be traded for the price $P_1 - P_0$ (or higher).

In contrast to a tax, a permit scheme does not impose as great an initial financial burden on the industry (assuming permits are given out, rather than sold). However, the incidence of the permit price will be like the incidence of a tax as permits are bought and sold among firms (the revenues from permits as outlined here would remain within the industry, whereas tax revenues presumably would accrue to the public treasury). For convenience, assume that world cadmium prices remain at P_0. A tax on cadmium in the same amount as the permit price $P_1 - P_0$ leads to the same allocation of cadmium use as does the permit system. Because of reduced use of cadmium, both the tax and the permit system would impose consumer surplus losses, in an amount given by the shaded triangle cde. The tax would impose an additional burden on the industry in the amount ($P_1 - P_0$) · Q^0, where Q^0 is the total permitted cadmium (assuming none of the tax revenues are returned to the industry). In contrast, the permit system results in the transfer of money between firms, but has no overall financial effect on the industry.

A Deposit-Refund System for Nickel–Cadmium Batteries

To recover cadmium from nickel–cadmium batteries, we suggest a modified deposit-refund system designed to promote the least-cost means for collection of batteries.[8] In a traditional deposit-refund system, a deposit is paid at the time of purchase and fully refunded

---

[8]Porter (1978), Bohm (1981), Bohm and Russell (1985), and Menell (1991) discuss the economics of deposit-refund systems. Erickson (1991) describes some of the current activity in battery recycling.

upon return of the product. Such a scheme provides two desirable incentives. First, the refund acts as a positive incentive to return the product. Second, for users who do not find it worthwhile to return the product, the deposit acts as a tax, serving to reduce demand and to encourage the use of substitutes. Although traditional deposit-refund mechanisms are often considered in association with recycling, this link is not essential. However, the return policy does make sense only when there is some alternative disposal practice that is clearly preferred to the existing disposal practice.

The traditional deposit-refund system has two disadvantages in its application to nickel–cadmium batteries. Reliance on the consumer to return batteries to the retailer or manufacturer may not be the least-cost means for collecting batteries. Further, given the disposal problems discussed previously, it seems appropriate that there should be some net tax to discourage overall use. If this is appropriate, then the deposit on batteries might be less than 100 percent fully refunded, even if 100 percent of the batteries were returned.

The deposit-refund could work as follows. A product tax on cadmium used in the manufacture of batteries would serve as the deposit. Refunds from these tax revenues would be paid out to collection enterprises, which might include any local government or private firm, with payment based upon the number of batteries collected and verification of their appropriate disposal. The collection enterprises would themselves decide how to best ensure the return of batteries. They might elect to compensate retail stores, or other local groups, or municipal waste collectors for their collection efforts.

The profit motive should drive the search for the least-cost means of collection. For example, suppose an incinerator plant finds that it can mechanically remove batteries from the waste stream at a modest cost. A collection enterprise may find that buying batteries from these plants is less costly than trying to encourage customer returns. (The collection enterprise and the waste plant might be the same entity.) Curbside pickup programs may similarly be found to be effective means for collection and, if so, would be supported by the collection enterprises. A retail store or manufacturer choosing to participate in the battery collection program would transfer the batteries to the collection enterprises and be compensated at a negotiated rate. The store or manufacturer, in turn, might find it effective to encourage consumer returns by refunding money to customers who return batteries, in the manner of the traditional deposit-refund system.

The incentive effects of this scheme are somewhat different than for the traditional deposit-refund mechanism. Usually the refund is targeted at the consumer and is intended to increase the incentive to return the product. There is, however, some doubt as to whether a refund in itself would be sufficient to encourage consumers to return an infrequently discarded product like nickel–cadmium batteries. Our scheme instead relies on the profit motive of collection enterprises in searching for the best way to recover batteries. The refund to consumers might remain as one component of this system, to the extent that it contributes to increased returns. For the consumer who chooses not to return batteries, the tax passed on in higher battery prices acts to reduce use and shift demand toward less toxic substitutes. (Ideally, other harmful batteries would be covered by similar collection programs, with taxes set by product type.)

The final disposition of the batteries could be either controlled storage, disposal, or sale to recycling plants. With the sharply fluctuating and often low prices for cadmium, the environmental concerns associated with recycling plants, and the costs of separating nickel–cadmium batteries from other types of batteries, there seems to be little incentive for recycling. As a result, government efforts to promote recycling might face the frequent need to subsidize the industry during periods of low prices. The likelihood of these subsidies, the expected high costs of a collection system, and the potential exposure problems associated with recycling make it doubtful whether recycling is preferable to a program of controlled disposal or storage.

REFERENCES

Anulf, T. 1989. "SAB NIFE Recycling Concept for Nickel–Cadmium Batteries: An Industrialized and Environmentally Safe Process." Pp. 161–163 in Sidney A. Hiscock and Rosalind A. Volpe, eds., *Cadmium 89.* Proceedings of the Sixth International Cadmium Conference, Paris, April 19–21 (London: Cadmium Association).

Bohm, Peter. 1981. *Deposit-Refund Systems: Theory and Application to Environmental, Conservation, and Consumer Policy* (Baltimore, MD: Johns Hopkins University Press for Resources for the Future).

————, and Clifford S. Russell. 1985. "Comparative Analysis of Alternative Policy Instruments." Pp. 395–433 in Allen V. Kneese and James L. Sweeney, eds., *Handbook of Natural Resources and Energy Economics* (New York: North-Holland).

Bromley, J., P. J. Young, P. Rushbrook, and J. Bentley. 1983. "Environmental Aspects of the Release and Fate of Cadmium in Municipal Landfills, with Reference to the Use and Disposal of Nickel–Cadmium Batteries and Pigmented Plastics." Pp. 61–66 in David Wilson and Rosalind A. Volpe, eds., *Cadmium 83*. Proceedings of the Fourth International Cadmium Conference, Munich, March 2–4 (London: Cadmium Association).

Cadmium Association. 1978a. *Technical Notes on Cadmium: Cadmium Coatings* (London).

———. 1978b. *Technical Notes on Cadmium: Cadmium in Alloys* (London).

———. 1978c. *Technical Notes on Cadmium: Cadmium in Stabilizers for Plastics* (London).

———. 1979. *Technical Notes on Cadmium: Cadmium in Batteries* (London).

———. 1980. *Technical Notes on Cadmium: Cadmium Production Properties and Uses* (London).

———. 1990. *Cadmium 1990: A Review* (London).

———. No date. *Cadmium Looks to the Future: Pigments* (London).

Erickson, Deborah. 1991. "Cadmium Charges: The Environmental Costs of Batteries Are Stacking Up." *Scientific American*, vol. 264, no. 5 (May), p. 122.

Goulder, Lawrence H. 1991. "Effects of Carbon Taxes in an Economy with Prior Distortions: An Intertemporal General Equilibrium Analysis for the U.S." Mimeo (Palo Alto, CA: Stanford University).

Hahn, Robert W., and Roger G. Noll. 1990. "Environmental Markets in the Year 2000." *Journal of Risk and Uncertainty*, vol. 3, no. 4 (December), pp. 351–367.

Hiscock, Sidney A., and Rosalind A. Volpe, eds. 1989. *Cadmium 89*. Proceedings of the Sixth International Cadmium Conference, Paris, April 19–21 (London: Cadmium Association).

Kelecom, H. 1989. "Factors Affecting Current Markets and the Outlook for Cadmium—A Metal Trader's Viewpoint." Pp. 7–10 in Sidney A. Hiscock and Rosalind A. Volpe, eds., *Cadmium 89*. Proceedings of the Sixth International Cadmium Conference, Paris, April 19–21 (London: Cadmium Association).

Kjallman, A. 1989. "The Effects and Experience of the Swedish Ban on Cadmium—Actions Concerning the Collection of Spent Nickel–Cadmium Batteries." Pp. 36–37 in Sidney A. Hiscock and Rosalind A. Volpe, eds., *Cadmium 89*. Proceedings of the Sixth International Cadmium Conference, Paris, April 19–21 (London: Cadmium Association).

Lauwerys, Robert R., and Dennis Malcolm. 1985. *Health Maintenance of Workers Exposed to Cadmium: A Guide for Physicians* (New York, NY: Cadmium Council, December).

McClure, S. 1989. "New Developments and Markets for Nickel–Cadmium Batteries in North America." Pp. 68–70 in Sidney A. Hiscock and Rosalind A. Volpe, eds., *Cadmium 89*. Proceedings of the Sixth International Cadmium Conference, Paris, April 19–21 (London: Cadmium Association).

Menell, Peter S. 1991. "Optimal Multi-tier Regulation: An Application to Municipal Solid Waste." Draft (University of California, Berkeley).

Oates, Wallace E. 1991. "Pollution Charges as a Source of Public Revenues." RFF Discussion Paper QE92-05 (Washington, DC: Resources for the Future).

Organisation for Economic Co-Operation and Development. 1991. "Co-Operation on Existing Chemicals: Risk Reduction, Lead Country Report on Cadmium." Paper presented at the 16th Joint Meeting of the Chemicals Group and Management Committee, Paris, May 28–30.

Porter, R. C. 1978. "A Social Benefit-Cost Analysis of Mandatory Deposits on Beverage Containers." *Journal of Environmental Economics and Management*, vol. 5 (December), pp. 351–375.

Shagarofsky-Tummers, A. 1989. "European Action Programme to Combat Environmental Pollution by Cadmium—Actions Underway and Outlook." Pp. 31–35 in Sidney A. Hiscock and Rosalind A. Volpe, eds., *Cadmium 89*. Proceedings of the Sixth International Cadmium Conference, Paris, April 19–21 (London: Cadmium Association).

U.S. Bureau of Mines. 1990. *Mineral Commodity Summaries 1990* (Washington, DC: U.S. Department of the Interior).

Volpe, Rosalind. 1985. "Environmental and Health Effects of Cadmium Pigments and Stabilizers in Plastics." Based on a paper presented at the Society of Plastics Engineers Annual Technical Conference (New York, NY: Cadmium Council).

Wilson, David, and Rosalind A. Volpe, eds. 1983. *Cadmium 83*. Proceedings of the Fourth International Cadmium Conference, Munich, March 2–4 (London: Cadmium Association).

———. 1986. *Cadmium 86*. Proceedings of the Fifth International Cadmium Conference, San Francisco, February 4–6 (London: Cadmium Association).

5

Brominated Flame Retardants

Brominated flame retardants (BFRs) are organic chemicals used as additives to reduce the flammability of plastics and textiles. BFRs present a special regulatory challenge because their use is largely driven by fire safety concerns; thus, they benefit private parties. However, production and disposal may pose environmental harm, thus imposing costs on third parties. Therefore, fire safety codes and other safety measures promoting the use of BFRs must be balanced with regulation to protect health and the environment.

Many brominated organic compounds can be used as flame retardants. Among related chemicals are some that have gained considerable notoriety in the past—for example, tris(2,3-dibromopropyl) phosphate (called Tris) and polybrominated biphenyls (PBBs). Tris was used as a flame retardant in children's sleepwear until it was found to be a potential carcinogen and mutagen that could be absorbed through the skin. PBBs, carcinogens analogous to polychlorinated biphenyls (PCBs), were widely used in plastics in the early 1970s until an accidental contamination of animal feed in Michigan forced the destruction of thousands of cattle and hogs and millions of chickens. Plastics manufacturers then rapidly switched to polybrominated biphenyl ethers (PPBEs) and other brominated or chlorinated compounds considered to be less toxic. However, recent concerns over the potential long-term risks associated with these and other BFRs now in common use have triggered yet another search for substitutes.

This history suggests several special characteristics of the flame-retardant industry in addition to the need to balance the goals of fire safety and environmental protection. First, there is a large and expanding number of BFR-related chemicals. At present, if the use of one BFR is restricted, then the most likely substitute is another brominated hydrocarbon. Second, knowledge is limited regarding the environmental fate and long-term toxicity of many of the compounds, particularly those that are new or that are used infrequently. Third, there is a considerable range of toxicity within the group of BFRs. Fourth, although there are many potential substitutes for BFRs (considerable industry attention is now directed toward phosphorus-based compounds), some compounds within most groups of substitutes have proven to be very toxic. Thus, it is not altogether clear that non-BFR substitutes will be less toxic. Nor is it clear that any new BFRs will be more toxic than non-BFR substitutes.

This case study considers the set of BFRs as a whole, with particular attention to the possibilities for substitution. As in the preceding chapters, we discuss production and use, potential health effects, pathways of exposure, and relevant legislation. We then consider the possibilities for incentive-based regulation, focusing on the key attributes of BFRs that influence the choice of regulatory approach. After commenting on the regulatory options, we propose a particular incentive-based mechanism and point out its strengths and shortcomings.

PRODUCTION AND USE

Production of Brominated Flame Retardants

Bromine is a heavy, red-brown liquid that produces irritating, corrosive fumes at room temperature. It is used in fumigants, pesticides, sanitizers, medicine, dyes and oil-well drilling fluids. Prior to the phasedown of lead in gasoline, the compound ethylene dibromide (EDB), which served as a lead scavenger, was the primary use of bromine. Now the primary use is in BFRs.

Although bromine was once extracted from seawater, almost all bromine now comes from concentrated underground brine deposits. Large deposits with high bromine concentrations are found in Arkansas and Michigan and in the Dead Sea area of Israel. The United States has virtually unlimited reserves. Worldwide bromine production in 1985 was estimated at 835 million pounds by the U.S. Bureau of Mines. The United States accounts for about 35 percent of world production, Israel for about 30 percent and the former USSR, the United Kingdom, and six other countries for the rest. U.S. production is around 300 million tons annually, down about 25 percent from peak production levels during the late 1970s when EDB usage was high. The United States exports bromine compounds as well as elemental bromine. Only small amounts are imported. It is estimated that as much as 35 percent of U.S. bromine may go to flame-retardant and other fire-safety uses (U.S. Bureau of Mines, 1990).

There are fewer than five U.S. producers of brominated flame retardants, with the industry dominated by the Great Lakes Chemical Company and the Ethyl Corporation.[1] Most production information is proprietary, although U.S. production has been estimated at approximately 80 million pounds per year. BFRs are produced by the bromination of hydrocarbons or oxygenated organic compounds. Often, more than 75 percent of the weight of the resulting compound is bromine.

There are two distinct types of BFRs: additive and reactive. The additive compounds are mixed into polymers, with no chemical reaction. The reactive compounds chemically react to become part of the polymer in which they are incorporated. Tetrabromobisphenol A (TBBPA) and its derivatives are the most frequently used of all BFRs. They are most commonly used as reactive retardants, although TBBPA can also be used as an additive flame retardant. Among the additive BFRs, decarbromobiphenyl ether (Deca) is the most heavily used. Other polybrominated biphenyl ethers (PBBEs)—octa- and pentabromobiphenyl ether—are also widely used. Numerous other additive and reactive BFRs are in use, including proprietary compounds known only by their trade names.[2]

The Market for BFRs

The use of flame retardants in plastics and textiles has been driven by the need to meet government regulations or, in some cases, voluntary industry standards for flame retardancy. About 90 percent of flame-retardant materials are used in plastics, particularly for the electronics, transportation, and building industries. BFRs are found in television sets and appliance cabinets, electrical wires, circuit boards, automobile plastics, seats and panels in planes and public transportation, and other uses where fire risks are high. Textiles (carpet and curtain backing), rubber, and paper products (for example, engine air filters) constitute the remaining uses. The number of firms incorporating flame retardants in their products is probably about 1,000.[3]

---

[1]Information on production is drawn from Chemical Products Corporation (1987).

[2]See Margler (1982), Siegman and coauthors (1988), Teurerstein (1988), and Szarek (1990).

[3]Information on the use and properties of flame retardants is drawn from Landrock (1983), Dick (1987), Radian Corporation (1987), Cullis (1988), and Troitzsch (1988).

Unfortunately, there is no ideal flame-retardant material for plastics. Some compromise is usually required between the physical properties and costs of the plastic, on the one hand, and the desired degree of flame retardancy, on the other. As a result, industry uses numerous flame retardants, each tailored to the nature of the polymer material and its end use. In addition, once having found a suitable flame retardant a company is generally reluctant to change, because the switch may alter the desired characteristics and manufacturing requirements of the product.

The flame retardants used in plastics fall into three main categories: inorganics, nonreactive organics, and reactive organics. Inorganics account for the largest proportion of flame retardants in use. They are cheap and suitable for use when a high degree of flame retardancy is not required and when the structural requirements of the plastics are not stringent. The large quantity of these materials that must be used in a typical product (30 to 40 percent by weight) tends to degrade some of the desirable properties of a plastic. Some inorganics are useful only as coadditives, that is, in conjunction with organic compounds. The inorganic compound alumina trihydrate is the most frequently used of all flame retardants. Toxic antimony oxides are the next most commonly used, but they are effective only when used in conjunction with halogenated organic flame retardants (they reduce the amount of the organic compound needed). Boron compounds, another group of frequently used inorganic flame retardants, are also coadditives that are used with the halogenated compounds.

Organic compounds are used when it is necessary to meet stringent flame-retardancy requirements. They include chlorinated, brominated, phosphorus, and halogenated phosphorus compounds. Brominated compounds are more effective—although more expensive—than chlorinated compounds, requiring that less material be added to the plastic. This smaller amount of BFRs (10 to 12 percent by weight) results in less degradation of the physical properties of the plastics. Phosphorus compounds are widely used in flexible polyvinyl chloride plastics, where they provide both flame resistance and plasticity. The phosphorus compounds are often combined with halogens, particularly in urethane foams, because the flame-retardancy effects of the two substances are complementary and the halogen increases water resistance. However, growing concern over the potential long-term risks of the halogenated flame retardants has spurred interest in finding phosphorus compounds that can serve as substitutes across a broader range of uses.

The choice among additive or reactive organic flame retardants is usually based on the expected impact on manufacturing costs. Reactive flame retardants essentially result in a new polymer, requiring considerable effort in formulating the plastic composition and in designing the production processes. Additive flame retardants are mixed in with an existing plastic, requiring less effort in reformulation. Reactive flame retardants have the advantage of minimal leaching or decomposition, whereas additive compounds tend to migrate and bleed, gradually losing their flame retardancy. Reactive flame retardants are used mainly in thermoset plastics (unsaturated polyesters, epoxy resins, and polyurethane foams) in which they can be easily incorporated. The additive flame retardants are widely used in thermoplastics (nylon, styrenes, polyethylenes, polypropylene, polyvinyl chloride, polycarbonates), but are also used in epoxies and urethane foams.

The choice of a particular flame retardant within a class depends upon a variety of factors, including its ability to withstand mixing; its ability to remain stable under processing temperatures, exposure to light, and conditions of use; and its aesthetic properties. Factors for additive compounds include the ability to remain well distributed in active form throughout the plastic. In all cases, cost and effectiveness figure as well. To be effective, a flame retardant must become active in the same temperature range at which the plastic would burn. Deca is often used in the high-impact polystyrene found in cabinets for household electrical appliances. It is stable under high temperatures, highly flame retardant, and its light color allows the plastic to be painted or colored. Penta, although somewhat less heat-tolerant than Deca, is useful in foams because it does not smolder. Widely used aromatic BFRs, such as PBBEs and TBBPA, have strong halogen–carbon bonds that withstand high processing temperatures as well as continued product use at high operating temperatures. The aliphatic BFRs have weaker bonds that tend to break under heat.

HEALTH EFFECTS AND PATHWAYS OF EXPOSURE

Except for BFRs that have been produced in large volume for several years, there is little solid information as to the environmental fate or long-term health effects of these flame retardants. Within the past few years, EPA has issued proposed test rules under section 4(a) of the Toxic Substances Control Act (TSCA) requiring the industry to determine the health effects and chemical fate of several BFRs (52 Fed. Reg. 25,219 [1987], 56 Fed. Reg. 29,140 [1991]). The testing requirements were triggered by evidence that significant releases of

BFRs are taking place, by indications that BFRs persist in soils and water sediments, and by concerns that BFRs may accumulate in animal tissue and rise through the food chain to pose human health risks. Some BFRs have been linked to immune suppression, developmental effects, and cancer in animals. There is, however, no specific evidence of human health effects resulting from releases of BFRs that are in current use.

Most BFRs are toxic when ingested, and a few can be absorbed through the skin, with toxic effects. This absorption can even occur after the flame retardant (Tris, for example) is incorporated into a polymer. Acute or short-term exposure to brominated compounds may damage the liver, kidneys, and thyroid. Among the widely used BFRs, Deca is considered to have low acute toxicity but was linked in one study to cancer of the pancreas in rats. No data on human cancers are available. In general, BFRs are believed to be considerably less toxic than chlorinated compounds. It is also thought that the risk of carcinogenicity decreases with increased bromination within a family of compounds (such as PBBs or PBBEs).[4]

Releases of BFRs to the environment can occur during initial production and in the manufacturing of plastics. Exposure during the use of products is apparently negligible, although some of the flame retardant is released from plastics at higher temperatures. Workers in both the flame-retardant and plastics industries can be exposed to dusts and some fumes, particularly during the packaging and unpacking of BFRs used in powdered form.

Product disposal may be a source of concern. Incineration of plastic containing BFRs should destroy most of the flame retardant. Under incineration, the brominated compounds generally form hydrobromic acid, an irritant in low concentrations but not a cause for general concern (although it is a risk to fire fighters exposed to smoke from burning plastic). However, incineration may also lead to the release of some highly toxic and persistent by-products such as brominated dioxin and furans (Buser, 1986). PBBEs have a greater tendency to produce these compounds. Direct disposal into landfills is

---

[4]Reviews or references to studies on the toxicity of the BFRs can be found in IARC publications on PBBs (World Health Organization, 1986) and Deca (World Health Organization, 1989), in the Interagency Test Committee reports that preceded the issuance of TSCA test requirements for BFRs (U.S. EPA [1989]), in a report by the Organisation for Economic Co-Operation and Development (OECD, 1991) and in Norris (1975). See Winzler and Kelly Consulting Engineers (1982) for an earlier EPA-sponsored review of environmental concerns with respect to the bromine-based chemical industry.

Table 5-1. Brominated Flame Retardant Life Cycle and
Sources of Exposure

Life-cycle stage	Exposure
Production of BFRs	Worker
	Ambient air and water
Production of plastics	Worker
	Ambient air, water, and soil
Consumption of plastics	Consumer
Disposal	Ambient air, water, and soil

also a potential source of releases, since some additive flame retardants slowly leach from plastics.

Production wastes and emissions are considered to be the most
important sources of environmental contamination (OECD, 1991;
Wong, 1986). During the production of BFRs, airborne dust and
waterborne emissions can spread BFRs to nearby soils and water. In
plastics manufacturing, the scrapings from equipment usually go into
landfills unless they are recycled, and they can be a source of flame-
retardant wastes. BFRs tend to decompose under exposure to light.
However, once emissions enter soils and stream sediments they may
only break down very slowly. Soil around production facilities has
been found to contain BFRs. There is also evidence of BFRs in stream
sediments, shellfish, and sea birds and mammals. Table 5-1 summarizes the life-cycle risks of BFRs.

EXISTING U.S. REGULATION OF BFRs

BFRs are not heavily regulated under current federal environmental
legislation. They are not listed as hazardous wastes under the Resource Conservation and Recovery Act (RCRA), or as air toxics under
the Clean Air Act, nor are they priority pollutants under the Clean
Water Act. Worker exposure to BFRs is not explicitly controlled,
although BFR dust would fall within nuisance particulate limits.
However, BFRs can be regulated under state implementation plans
for the Clean Air and Water acts. The commonly used Deca is on
the Superfund Amendments and Reauthorization Act (Community
Right-to-Know Act, Title III, Emergency Planning and Provisions)
list of toxic chemicals, which requires that its environmental releases
be reported. Finally, as noted above, testing to determine the health
effects and chemical fate of several BFRs is now required under
section 4(a) of TSCA.

Perhaps of greatest importance to the use of BFRs have been the requirements of limiting flammability in plastics. These requirements include a broad range of regulations, state and local codes, and voluntary industry standards that drive the demand for highly effective flame retardants. Plastics in the electrical and electronics industry meet standards set by Underwriters' Laboratory, Inc., and the International Electrotechnical Commission. Federal regulations related to the flammability of plastics include the U.S. Federal Motor Vehicle Safety Standard 302 and the U.S. Federal Aviation Regulations on Airworthiness. Also of some importance has been the reduced demand for bromine that followed the EPA-mandated phasedown of lead in gasoline and the restriction on use of EDB as an agricultural fumigant. Reduced demand for bromine tended to reduce its price, making the use of brominated flame retardants more attractive to plastics producers.

KEY ATTRIBUTES OF REGULATORY INTEREST

This section summarizes some characteristics of BFRs and the flame-retardancy market that are relevant to the choice of incentive-based regulations.[5]

BFRs are often toxic, and it is suspected that many persist in the environment and cause long-term human health effects. Emissions during production are a primary source of environmental contamination, although incineration of BFR-containing products can result in the release of very toxic decomposition products. Gradual leaching of additive flame retardants can occur from plastics in landfills.

Another important characteristic, described earlier, is that BFRs are embodied in plastic products, and flame retardants in plastics alter the physical properties of the plastics in which they are incorporated, significantly complicating the process of switching to substitute flame retardants. Another key attribute of BFRs is that toxic exposure during product use, while not impossible, is negligible for commonly used compounds.

The recovery of BFRs from plastics is likely to be impractical, although these flame retardants do not greatly change the recyclability of a plastic. The separation of plastics containing BFRs to ensure

---

[5]See OECD (1991) for an alternative overview of the BFRs and related regulatory concerns.

that they are incinerated under controlled conditions is likely to be moderately expensive. At present, the identification of plastics containing particular flame retardants is difficult, since no identifying markings are currently required. Chemical incineration to stop the release of toxic decomposition products would probably be very costly. Removing BFR-containing plastics from the waste stream going to landfills might be impractical, given potentially high costs of separation and the lack of storage options.

Also important for considering regulatory options is that there are very few producers of BFRs, making it relatively easy to monitor plant emissions. There are numerous producers of plastics and the other polymer products that contain BFRs, however. Import volumes are negligible, although some flame retardants are embodied in plastic casings and circuit boards of imported electrical equipment.

There are various substitutes for halogenated flame retardants, but many of them are also toxic. In the less demanding applications, the switch to less toxic substitutes may be straightforward (although in many of these cases BFRs are probably less used now, because of their cost). In other applications a considerable development effort might be required to find a suitable flame retardant and plastic mix. Phosphorus-based compounds are coming to be viewed as the most likely replacement for halogenated flame retardants, but some phosphorus compounds are quite toxic, although they are not as environmentally persistent as some BFRs are. The plastics industry is making some effort to reduce use of BFRs, even without regulation.

This discussion indicates the desirability of innovation in the design of flame-resistant plastics to accommodate both the benefits of reduced flammability in BFR-containing products and any environmental risk associated with the BFRs. Because the flame-retardancy industry appears to have a host of possible directions for product innovation, a major concern in designing an incentive scheme to regulate the industry should be that it not discourage the introduction of new environmentally sound products.

POSSIBLE INCENTIVE-BASED INTERVENTIONS

As the preceding discussion indicates, the primary regulatory concerns with respect to BFRs are with their release from production facilities and with the disposal of BFR-containing plastics, which might lead to the release of BFRs and related products through decomposition. An ideal incentive-based mechanism would discourage plant emissions, discourage use of the more toxic flame retardants, encourage the introduction and use of less toxic substitutes, encourage the design of plastics that minimize leaching of flame retardants, and encourage the choice of appropriate means of disposal. Some of these goals—for example, discouraging plant emissions—could be met fairly easily. Others are likely to be more difficult to accomplish, or at least to accomplish simultaneously. For example, it would be easy to discourage the use of BFRs by taxing them, but it is far from clear whether such a strategy would encourage the use or introduction of less toxic alternatives. As in the case of chlorinated solvents (see chapter 2), a difficulty in designing a complete incentive mechanism to deal with flame retardants comes from uncertainty as to the relative hazards of BFRs and their potential non-BFR substitutes, particularly when the substitutes may be new products.

Plant Emissions. Controlling emissions from plants producing BFRs may be relatively easy, given the small number of plants. There are only a few sites where BFRs are produced, and even fewer firms. Primary emissions are found in waste water and airborne dust. With so few plants, and fewer firms, a tax based on measured or estimated emissions might be easy to handle administratively.

Disposal Practices. The goal in controlling the flow of BFR wastes should be to ensure that incinerated plastics are burned under controlled conditions to limit the release of dioxins. Limiting landfill disposal does not seem required at this time, as there is no practical alternative short of reducing the use of BFRs. Sorting BFR-containing plastics from landfill material would be costly because many are built into large products (for example, housings of electrical products, construction and demolition debris that includes wires and panels from commercial office buildings).

If sorting of wastes were to be required, an initial step might be a standardized marking system to show the presence and type of flame retardants. The additional costs of sorting should, in principle, also be reflected in the price of the product, to reflect the full social costs of product use. This could be accomplished through a tax on plastics containing any flame retardants that require special waste treatment. In contrast to the case of cadmium, there seems to be less gain from a deposit-refund system, given that more centralized sorting of collected wastes seems likely to be cheaper than sorting by the consumer.

Use of Flame Retardants. Taxes on BFR compounds would tend to discourage their use (and so disposal) and encourage the use of substitutes. However, depending upon its design, a tax could encourage the use of substitute flame retardants that are no less toxic than current BFRs, and the tax could discourage innovation in the design of new flame retardants.

Piecemeal taxation of flame-retardant compounds, in the absence of assurance that the alternatives are less toxic, could lead to undesirable consequences. Suppose that only the dominant BFRs were taxed. Usage would then switch to substitute materials, probably other brominated compounds, about which less is known but which might eventually prove to be of equal or greater toxicity.

A blanket tax on all brominated (or halogenated) flame retardants might be just as unacceptable. Such a tax would encourage a switch to substitute materials, but the nonbrominated compounds can themselves be considerably toxic. With a tax on all BFRs, there would be no incentive for users to switch to less hazardous BFR variants. Further, if the tax were applied to new products, it could effectively limit any incentive to develop new BFRs with reduced toxicity. Alternatively, exempting each new BFR from the tax for a limited "grace period" could encourage rapid introduction of new compounds simply as a means to avoid taxation.

A more idealized scheme that taxes each flame-retardant material on the basis of its expected environmental damages would avoid these problems of inappropriate incentives but could not be designed easily, given the lack of information on such damages. An effective incentive scheme in a situation characterized by considerable uncertainty about new products must be adaptive, adjusting to reflect new knowledge on safety while providing incentives for development of safer products.

PROPOSED INCENTIVE MECHANISMS

One promising approach to dealing with third-party damages associated with the releases of BFRs is a deposit-refund or performance-bond scheme targeted at the producers of the flame retardant. In addition, plastics containing flame retardants might be coded to show the type of material used. Flammability standards might also be reviewed to see if perhaps they unintentionally encourage use of the halogenated flame retardants. In table 5-2, various regulatory options are summarized, with brief comments.

Table 5-2. Regulatory Options for Brominated Flame Retardants

Exposure stage	Regulatory option	Verdict	Comments
Manufacture	Content labeling	No	May offer only limited options for changing response.
	Safety labeling	No	No user risks involved.
	Command-and-control ban	No	Imposes loss of user benefits, moderate industry cost; is uncertain as to risks from substitutes.
	Tradable permits	No	Are impractical, given the wide range of substitutes.
	BFR input tax	Maybe	May be difficult to estimate damages; may be difficult to ensure that substitutes are less risky.
	Producer deposit-refund	Yes/maybe	Shifts incentives for emission control to producer; shifts incentive to resolve uncertainties to producer; may be costly to administer system and difficult to determine appropriate deposits and refunds.
Product use	Product sales tax	No	Would be administratively easier to tax inputs of flame-retardants.
	Safety labeling	No	Few user risks involved.
	Review fire regulation	Yes	May be desirable to review whether the use of toxic flame retardants is driven by fire standards that are too strict.
Product disposal	Disposal fees	No	Would be impractical to target BFR products; offers little opportunity to alter disposal practice; may increase incentive for illegal disposal.
	User deposit-refund	No	Offers little opportunity to alter disposal practice.

Producer Deposit-Refund and Performance-Bond Systems

For this case, in which new BFR products may play an important role, where there is considerable uncertainty as to product damages, and where these damages may occur well after the period of production, producer deposit-refund systems can be valuable tools. Such generalized deposit-refund systems can move the incentives for ensuring product safety from the regulator to the producer. Similarly, these systems can shift much of the burden of proof from the public to the producer in resolving uncertainty as to damages associated with new products. The systems may also be designed to avoid some of the difficulties that can be associated with assessing ex-post liability for long-term damages linked to persistent chemicals.[6]

Under a producer deposit-refund scheme, manufacturers of BFRs and related flame retardants would be required to make deposits based on the anticipated environmental or health effects of their production. The deposits would be refunded at some future time, depending on the extent to which the producer acts to reduce damages or to demonstrate product safety.

The initial deposit rate would be based on estimates of the present value of marginal damages expected under unregulated production. The actual level for this deposit could be negotiated, with lower deposits called for when a firm provides convincing evidence of new product safety. Further deposits would be paid periodically, based on current production levels, with the rate for new deposits adjusted to reflect product-safety knowledge gained from continuing production and use, health studies, and evidence of the manufacturer's care in controlling emissions. The pool of deposits would be held in an interest-bearing account.[7]

Deposits would be returned at periodic intervals and could be based on the meeting of clearly defined conditions as to the control of releases, results of health studies, and the decay rate of the product. In a simple approach, the refund would occur if certain negotiated conditions had been met over the time interval. For example, a producer might be eligible for a full refund of the previous year's deposit if BFR levels measured in nearby soils and streams had been held below some agreed-upon level. This approach makes the mechanism for refunds well defined, but the overall success of such an

---

[6]See Ringleb and Wiggins (1990) for a discussion of liability and long-term hazards.

[7]Compound interest on deposits, even from a low initial rate of deposits, might prove sufficient to cover very high damages that occur after a long latency period.

approach relies upon a good correlation between meeting the measured criteria and the desired reduction in third-party damages.

Under a more ambitious approach, the regulator might attempt to maintain a pool of deposits sufficient to cover fully future health and environmental damages. This would correspond in principle to the approach suggested by Costanza and Perrings (1990). The deposit on hand would then cover the present value of the potential risk from both extant products and potential emissions as well as from current output. Refunds from past deposits would be paid back to the extent that the existing deposit pool exceeded the amount of anticipated damages. Evidence from health studies and product experience presented by the producer would provide the basis for gradual refinement in the estimates of anticipated product damages.

In some cases, firms might find themselves able to insure against the loss of deposits. For a premium, an insurance or bonding company could assume liability for the deposit and rights to the refunds. The producing firm could then post a performance bond with the government, guaranteeing insurance up to the amount of the required deposit (see Bohm and Russell [1985], Costanza and Perrings [1990]). Of course, an insurance company would take on such a liability only if it was reasonably sure of the product's safety. The insurer would also then assume an interest in seeing that the manufacturer took all appropriate steps to limit emissions and ensure product safety. The producer would find it advantageous to act to ensure product safety and emissions control, because these steps would reduce future insurance premiums.

The primary advantage of the producer deposit-refund or performance-bond system is that it gives incentives to the producer for ensuring product safety (not unlike product liability law, but in the case of deposit-refund or bonds, the liability link is perhaps more explicit ex ante). Both the return of current deposits and the level of deposits on future production are made to depend upon the producer's control over plant emissions and the demonstrated safety of product use and disposal. In addition, under these schemes, new products may be introduced at relatively low costs, especially when there is convincing evidence of safety. Since deposits are based on output levels, deposits are less of a hurdle to new-product entry than are high fixed costs of mandatory product testing. On the other hand, once there proves to be a potential for large sales, the producers themselves then have incentives to act to demonstrate and ensure product safety.

Under the producer deposit-refund system, uncertainty as to damages associated with new products is dealt with by the adaptive manner in which deposits or premiums are set and by the association of refunds with measurable performance. The risks associated with these uncertainties are largely shifted to the producers. Because of limited information, difficulties in the estimation of damages would remain, however; thus, periodic administrative reviews would be necessary to resolve the required level of deposits. Particularly with long-lived products, it could be hard for firms to establish the absence of potential damages. A less ambitious program that links deposits and refunds to the meeting of very specific and measurable criteria would seem more practical than the alternative in which the deposits are intended to fully cover estimated damages.

And, finally, the link between the pool of deposits and the actual liability for damages is also of concern. Costanza and Perrings (1990) envision the pool of deposits as a source of funds for the remediation of environmental damages or for compensation to injured parties. A key question, however, is whether the easy availability of a pool of deposits might encourage excessive liability judgments in the repair of environmental damages.

REFERENCES

Bohm, Peter, and Clifford S. Russell. 1985. "Comparative Analysis of Alternative Policy Instruments." Pp. 395–433 in Allen V. Kneese and James L. Sweeney, eds., *Handbook of Natural Resources and Energy Economics* (New York, NY: North-Holland).

Buser, Hans-Rudolf. 1986. "Polybrominated Dibenzofurans and Dibenzo-*p*-dioxins: Thermal Reaction Products of Polybrominated Diphenyl Ether Flame Retardants." *Journal of Environmental Science and Technology*, vol. 20, no. 4, pp. 404–408.

Costanza, Robert, and Charles Perrings. 1990. "A Flexible Assurance Bonding System for Improved Environmental Management." *Ecological Economics*, vol. 2, no. 1 (April), pp. 57–75.

Chemical Products Corporation. 1987. "Chemical Products Synopsis: Bromine" (Cortland, NY).

Cullis, C. F. 1988. "The Use of Bromine Compounds as Flame Retardants." Pp. 301–331 in D. Price, B. Iddon, and B. J. Wakefield, eds., *Bromine Compounds: Chemistry and Applications* (Amsterdam: Elsevier).

Dick, John S. 1987. *Compounding Materials for the Polymer Industries: A Concise Guide to Polymers, Rubbers, Adhesives, and Coatings* (Park Ridge, NJ: Noyes Publications).

Landrock, Arthur H. 1983. *Handbook of Plastics Flammability and Combustion Toxicology: Principles, Materials, Testing, Safety, and Smoke Inhalation Effects* (Park Ridge, NJ: Noyes Publications).

Margler, Lawrence W. 1982. "Project Summary: Environmental Implications of Changes in the Brominated Chemicals Industry," EPA-600/S8-82-020 (Washington, DC: U.S. Government Printing Office, September).

Norris, J. M., et al. 1975. "Toxicology of Octabromobiphenyl and Decabromobiphenyl Oxide." *Environmental Health Perspectives*, vol. 11 (June), pp. 153–161.

OECD (Organisation for Economic Co-Operation and Development). 1991. "Co-Operation on Existing Chemicals: Risk Reduction, Lead Country Report on Brominated Flame Retardants." Paper presented at the 16th Joint Meeting of the Chemicals Group and Management Committee, Paris, May 28–30.

Radian Corporation. 1987. *Chemical Additives for the Plastics Industry: Properties, Applications, Toxicologies* (Park Ridge, NJ: Noyes Data Corporation).

Ringleb, A. H., and S. N. Wiggins. 1990. "Liability and Large-Scale, Long-Term Hazards." *Journal of Political Economy*, vol. 98, no. 3, pp. 574–595.

Siegman, A., S. Yanai, A. Dagan, Y. Cohen, M. Rumack, and P. Georlette. 1988. "Poly (Pentabromobenzyl Acrylate). A Novel Flame-Retardant Additive for Engineering Thermoplastics." Pp. 339–351 in D. Price, B. Iddon, and B. J. Wakefield, eds., *Bromine Compounds: Chemistry and Applications* (Amsterdam: Elsevier).

Szarek, Pat. 1990. "Economic Analysis of Proposed Test Rule for Five Brominated Flame Retardants, Non-CBI Version." Internal memo (Washington, DC: U.S. Environmental Protection Agency, Office of Pesticides and Toxic Substances), October.

Teurerstein, A., R. Scharia, M. Rumack, S. Yanai. 1988. "New Flame Retardant Additives for V-2 Polypropylene." Pp. 332–338 in D. Price, B. Iddon, and B. J. Wakefield, eds. *Bromine Compounds: Chemistry and Applications* (Amsterdam: Elsevier).

Troitzsch, H. J. 1988. "Flame Retardants." Pp. 709–747 in R. Gachter and H. Muller, eds., *Plastics Additives Handbook*, 3rd ed. (New York, NY: Hanser).

U.S. Bureau of Mines. 1990. *Mineral Commodity Summaries 1990* (Washington, DC: U.S. Department of the Interior).

U.S. EPA (Environmental Protection Agency). 1989. "Twenty-fifth Report of the Interagency Testing Committee to the Administrator; Receipt of Report and Request for Comments Regarding Priority List of Chemicals." *Federal Register*, vol. 54, no. 237 (Dec. 12), pp. 51,114–51,134.

Winzler and Kelly Consulting Engineers. 1982. *Environmental Implications of Changes in the Brominated Chemicals Industry*. Final Report submitted to the U.S. Environmental Protection Agency (Eureka, CA April 7).

Wong, Kin F. 1986. "Production/Exposure Profile: Brominated Diphenyl Oxide." Mimeo, January 7. (Washington, DC: U.S. Environmental Protection Agency).

World Health Organization. 1986. *IARC Monographs on the Evaluation of Carcinogenic Risks to Humans: Some Halogenated Hydrocarbons and Pesticide Exposures*, vol. 42 (New York, NY).

World Health Organization. 1989. *IARC Monographs on the Evaluation of Carcinogenic Risks to Humans: Some Flame Retardants and Textile Chemicals, and Exposures in the Textile Manufacturing Industry*, vol. 48 (New York, NY).

6

Summary and Conclusions

The preceding chapters have considered the use of economic incentives as alternatives to command-and-control regulation of the undesirable health or environmental effects that might be associated with the production and use of chemicals. Our discussion has sought to tailor general prescriptions from the economics literature on environmental regulation to accommodate the special characteristics of toxic substances. The general literature typically assumes a fairly homogeneous pollutant associated with one stage of production (or use) at a somewhat readily identifiable source. The following characteristics of toxic substances challenge these assumptions, however:

1. The potential for health and environmental risks to occur at multiple stages of the life cycle of the substance;
2. Significant variation in the distribution of potential risks across a multitude of heterogeneous products and applications; and
3. The potential for exposure to other hazardous substances or processes that may be substituted for a regulated substance.

The first two of these factors call for regulatory interventions that target specific life-cycle stages, products, and uses. However, targeting intervention may be costly to administer, enforce, and monitor. Blunter interventions, such as an outright ban on a substance or a tax on all production or use of that substance, may be less costly to administer but may impose large social welfare losses if productive yet relatively harmless applications are precluded. Self-enforcing strategies such as deposit-refund mechanisms or, in the case of taxes and tradable permits, the use of audit/penalty schemes to induce compliance have the potential to reduce administrative, enforcement, and monitoring costs. (Audit/penalty schemes, not detailed in the preceding chapters, are outlined in Russell, Harrington, and Vaughan [1986].)

The third factor—substitution—calls for a more "general-equilibrium" approach to intervention, that is, an approach that incorporates information about possible substitutes for toxic substances and/or processes and the potential exposure risks associated with them. Such an approach imposes extensive information requirements, however. The posting of performance bonds may offer a way to encourage the use and development of safe substitutes, and we consider the analysis of bonds in greater depth to be a prime candidate for future research.

Our analysis also considers the regulation of substances (or products) for which third-party effects are negligible—that is, where the risks are to individual users rather than to society as a whole. Here we focus on product labeling to indicate the risk associated with one product in comparison with alternatives, and to suggest protective measures when using the product.[1] We also discuss the possible role that voluntary or mandated product standards might play in providing additional information to the user, although we do not consider in detail the desirability of voluntary versus required standards, nor do we investigate the likelihood of voluntary standards being adopted by industry.[2] We do note, however, the literature that indicates that labeling may induce meritless litigation. Each of these topics is also a candidate for additional research.

Specific conclusions derived in the preceding case studies of chlorinated solvents, formaldehyde, cadmium, and brominated flame retardants (BFRs) include the following prescriptions for incentive-based approaches. (As emphasized in the Introduction, in each case study we assume that regulators have deemed the potential exposure risk to be significant enough to warrant intervention.)

• *Chlorinated Solvents:* A deposit-refund scheme for residual solvent used in cleaning applications; labeling to inform consumers or workers of the potential hazards associated with exposure in applications where third-party effects are small; and a product tax imposed on the manufacturer for sales of solvent in dissipative applications.

• *Formaldehyde:* A combination of a tax on sales of formaldehyde at intermediate stages of production, to internalize third-party effects

---

[1]Labeling might also play a role in the externality (third-party) case if producers and consumers were willing to bear the cost of action taken for the greater social good.

[2]The extent to which workers are able and willing to respond to labels (for example, obtain protective masks, switch jobs) is an extant research issue in the literature. Workers may be imperfectly mobile, for example. For some evidence that workers do respond to some on-the-job risks, see Viscusi and O'Connor (1984).

arising from emissions during production; and labeling, to inform consumers of the potential for exposure in the home or workplace.

• *Cadmium:* A deposit-refund scheme for nickel–cadmium batteries, and a tradable permit mechanism for phasing out the use of cadmium in other cadmium-containing products.

• *Brominated Flame Retardants:* A product labeling scheme to indicate the presence of BFRs to permit controlled disposal, and performance bonds to encourage the substitution of safer substitutes, including new substances.

GENERALIZABILITY OF THIS ANALYSIS

Figure 6-1 offers a schematic overview of the life cycle, the market characteristics and regulatory setting, and the incentive strategies addressed in our case studies, indicating where the various case studies arise as examples. For instance, if an exposure potentially harmful to health or the environment arises during production (top panel of the figure), then possible incentive strategies include the approaches in the boxes in the bottom panel (tradable permits, taxes, contingent fees, and improved information). The applicability and effectiveness of these approaches depend, however, on factors suggested in the center panel, including market structure (number of producers, nature of substitutes for the product) and existing legislation and regulation.

It is probable that toxic substances other than those considered in this study can be slotted into the figure on the basis of an analysis of similarities between their characteristics and those of our case examples. Although such an exercise would offer a hint as to where in the life cycle intervention might be desirable, we emphasize that proposing an appropriate regulatory approach for different substances, even if they have in common the life-cycle stages where harmful exposure is a concern, is only at the level of suggestion.

Toward this end, figure 6-2 outlines a highly stylized "regulator's decision tree" that identifies various incentive-based regulatory approaches. The decision tree begins with ascertaining the possibility of significant toxic exposure; given a "yes" response, the tree branches to determining if exposure has third-party effects. If not, the tree branches to a determination of whether exposure is to workers or users and then to a series of steps that might be taken to improve (a) the supply of information provided to these individuals and (b) these individuals' demand for this information. The tree then indicates that inadequacies in information demand, supply, or both may

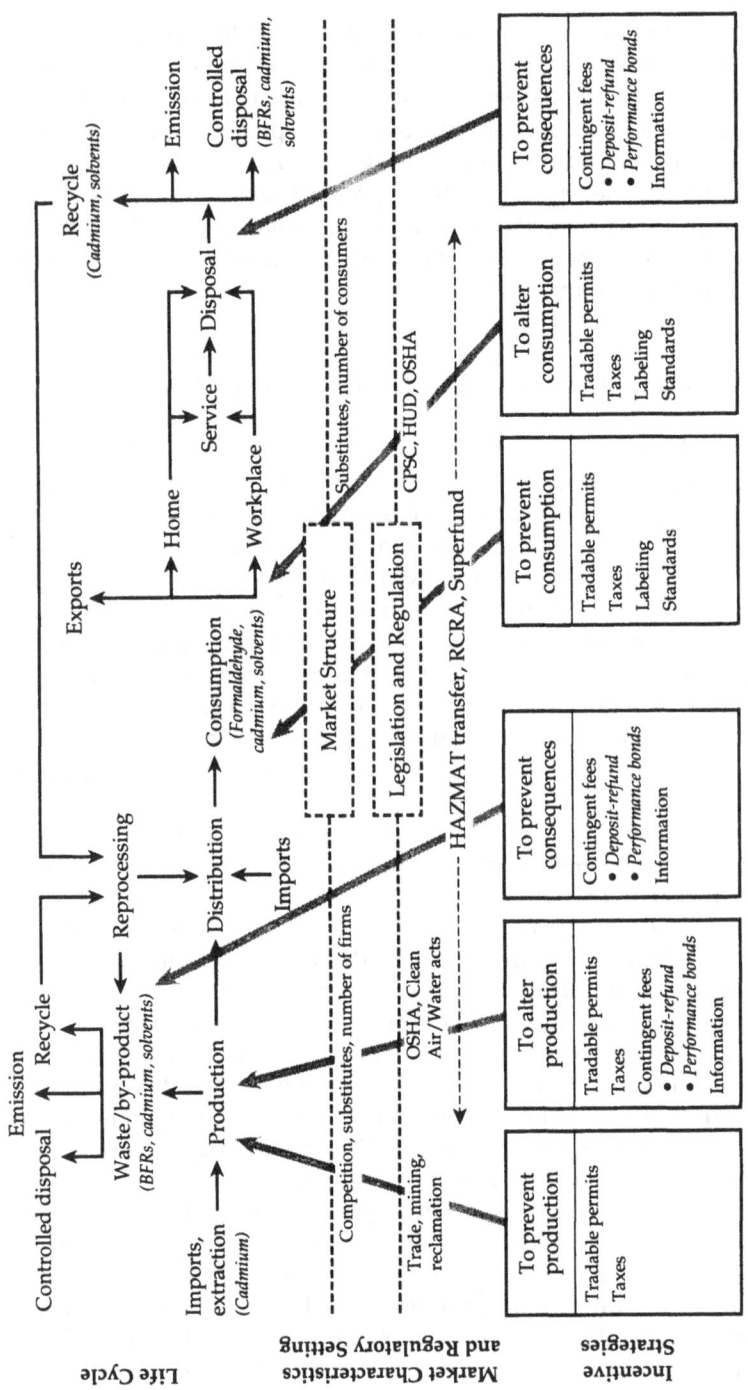

Figure 6-1. Economic incentives and the regulation of toxic chemicals: examples of exposure pathways and management interventions (see discussion in text).

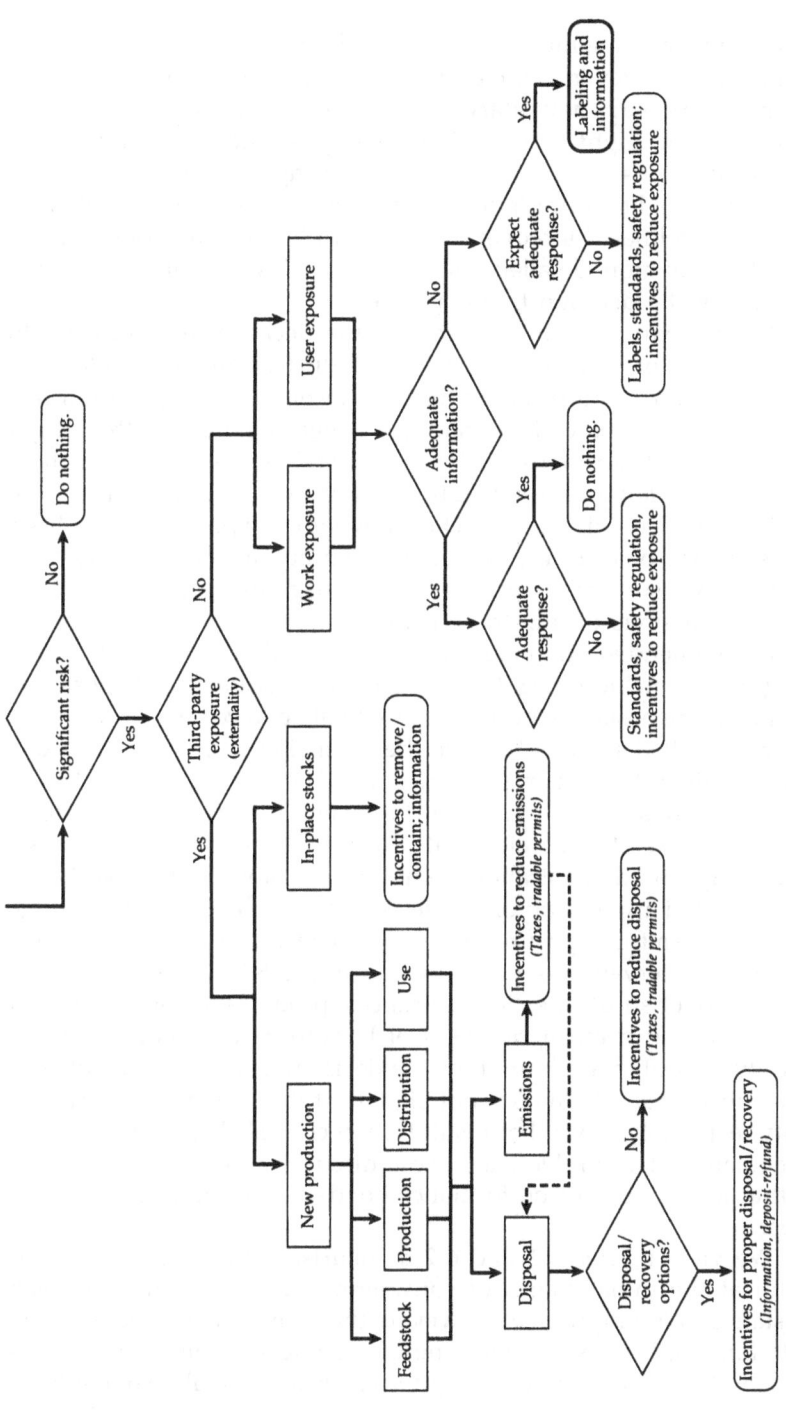

Figure 6-2. Regulator's decision tree (see discussion in text).

require labeling or some type of standard setting. What the tree does not indicate, however, is the precise form and extent of the mandated actions (or perhaps voluntary actions taken in response to "moral suasion") that might be desirable. Clearly, as indicated by the previous discussions on labeling, standard setting, and product liability, the economics literature is itself somewhat inconclusive in offering specific guidelines. Moreover, our case studies also demonstrate the importance of more detailed case-by-case analysis of substances once this part of the decision tree is reached.

If third-party effects are present, the decision tree branches to consideration of new versus existing products. Our case studies did not offer us the opportunity to focus on in-place stocks (for example, removal of asbestos or lead paint); we merely note that this might be a topic for future research on incentive-based strategies such as credits for removing or mitigating exposure. Regarding new production, the decision tree branches to whether exposure occurs from the feedstock or during production, distribution, or use, and then whether exposure arises primarily with disposal operations or with fugitive emissions during these activities.

Recommended strategies at the ends of the "new production" in figure 6-2 include incentives to reduce emissions or disposal, or to improve disposal, including the possibility of recovery for recycling or for controlled storage. As with the other side of the tree, the exact nature of the intervention strategy depends on a more detailed analysis of the substance in question (for example, is recycling feasible? Do markets exist for recycled material? Can taxes or permits be cost-effective given administration, monitoring, enforcement costs?).

Repeated application of the decision tree for each life-cycle stage at which harmful exposure arises would result in a set of possible regulatory interventions (according to figure 6-1). In addition, our concern about substitutes for substances, products, or processes that might result in even higher risks of harmful effects requires, theoretically, repeated application of the decision tree for the set of possible substitutes. Of course, not all substitutes are known with certainty, especially those that result from technical change in response to regulation. This problem led to our discussion in previous chapters on the possible use of performance bonds to encourage safe substitutes.

We view figures 6-1 and 6-2 as heuristics that are likely to be useful at the initial stages of considering how to regulate a given substance. We emphasize, however, that these generalizations are quite limited, because (1) tailoring of the intervention strategies implied by the figures is necessary to match individual characteristics of toxic substances;[3] and (2) the figures must be considered in a broad context that includes a variety of products and uses associated with a given substance (for example, the "parent" substance and the "child" products) as well as substitutes for the substance.

FUTURE RESEARCH

The preceding discussion suggests several directions for future research. As noted, extensions of the economics literature to consider performance bonds and product labeling in greater detail could prove quite useful, since these approaches seem to offer significant promise as regulatory mechanisms for toxic substances. In the case of performance bonds, specifically of interest is whether information about product risk, especially that associated with newly introduced products, is asymmetrically distributed between regulator and regulatee or if risk occurs only after time. Performance bonds may provide incentives to increase information collection (for example, encourage product testing) and information supply and, in particular, make less likely the substitution of a riskier product for the product presently regulated. Among the research questions about product labeling are the circumstances under which producers will voluntarily label and what types of labels elicit the most effective consumer responses.

The discussion above about generalizations that might be drawn from our study suggests that additional case studies would permit further refinement of the types of general rules of thumb that are suggested in figures 6-1 and 6-2, where the intent is to present guidelines that might aid regulators of toxic substances. Toward this end, additional research might include the specification in questionnaire format of information that regulators might need to design incentive-based techniques. Such information might include number of firms and consumers, including intermediate producers, at each stage of the life cycle, which is relevant to administrative costs; and the nature of secondary markets for substances or products for which deposit-refund and subsequent recycling might be a desirable alternative.

A simple cost–benefit model that makes explicit the tradeoff between intervention targeted at a specific use and the costs of monitoring and enforcing also might be useful, especially if it could be

---

[3]Industrial Economics, Inc. (1989), reaches a similar conclusion in a study of groups of chemicals regulated under Section 313 of SARA.

parameterized (empirically for a sample of substances, or perhaps by way of simulation) to derive rough benchmark numbers as guidelines for regulation. For instance, administrative costs might be hypothesized to increase with the number of firms at the relevant life-cycle stage, and an approximation of the number of firms beyond which administrative costs are likely to be too large to justify targeted intervention might be derived. A more formal cost–benefit analysis might also indicate the size of social savings from incentive-based approaches as well as winners and losers (for example, regional or industry employment impacts) that might be necessary to anticipate in order to make incentive approaches politically feasible.

Another topic, treated only lightly here, is the interactive effects of multiple layers of regulation of a substance. This topic includes possible effects of intervention at multiple stages of the life cycle (production, disposal; or workplace, home), as well as effects of intervention for different media (air, water, soil) or other factors (hazardous materials transportation). Without analysis of how these effects combine with one another—whether, when taken together, they offset or reinforce desired effects—we do not know if our proposed intervention strategies are, on net, welfare enhancing or whether, when considered together, they might push a product out of the market. Accordingly, a future research topic might consider interactive effects in a few case studies to test whether these distortions are worth further scrutiny by regulators.

Another important extension of this research would be to refine the definition of cost-effective regulation. This would involve consideration of how a particular regulatory approach affects the future technology choices of firms, particularly regarding the relative health and environmental effects of substitute processes or products. The research could consider whether regulatory interventions appropriate under a static concept of efficiency would still be appropriate for achieving dynamic efficiency. An additional consideration would be whether there are reasons (perhaps in terms of signaling intent to regulate in the future) why command-and-control regulatory approaches might have more desirable dynamic features.

REFERENCES

Industrial Economics, Inc. 1989. "Multi-Media Incentives for Reducing Release of SARA 313 Chemicals" Draft (Cambridge, MA, September).

Russell, Clifford S., Winston Harrington, and William J. Vaughan. 1986. *Enforcing Pollution Control Laws* (Washington, DC: Resources for the Future).

Viscusi, W. Kip, and Charles J. O'Connor. 1984. "Adaptive Responses to Chemical Labeling: Are Workers Bayesian Decision Makers?" *American Economic Review*, vol. 74, no. 5 (December), pp. 942–956

Index